THE
RULES OF
INNOVATION

BART
HUTHWAITE

INSTITUTE FOR LEAN INNOVATION
MACKINAC ISLAND, MICHIGAN
2007

Published by
Institute for Lean Innovation
P.O. Box 1999
Mackinac Island, Michigan

906-847-6094
www.InnovationCube.com

ISBN 13: 978-0-9712210-4-8

Cover and interior design and layout by Lee Lewis Walsh,
Words Plus Design,www.wordsplusdesign.com

Manufactured in the United States of America

Dedication

This book is dedicated to our five grandchildren, who will be the best to judge whether the innovations of today have brought happiness to their lives of tomorrow.

Kate Huthwaite
Charles Barton Huthwaite

Caroline Martin
Aidan Martin
Avery Martin

CONTENTS

Acknowledgements..ix

Preface..xi

Introduction..1

Innovation as the ultimate leadership skill... Lessons from the Model T... A short "Rules of Innovation" glossary... What you will learn in this book

Part I – How to Find Your Best Opportunities.............................13

Chapter 1 – Recognize Ripening Bananas 15

The best competition is no competition... Why you must detect opportunities and threats while they are still green... Growth opportunities are not created; they are discovered and exploited... All new *Things*, or solutions, come from recombining existing knowledge in new ways... The three forces of change keep the doorway of opportunity constantly open... How to recognize these "sharks" and leverage them to your advantage

Chapter 2 – Think Backwards..33

Start with the end-in-view... Why you must climb your mountain in reverse... How to apply "effect-to-cause" thinking... Why identical innovations occur at the same time...   Importance of speed-to-market.... Why a stakeholder who is not part of the solution will become part of the problem... How to get all stakeholders climbing your same mountain... Why your first ideas are never the best... How to move back and forth between opportunities and solutions to find the best path up your mountain

Chapter 3 – Explore for Hidden Wants 50

Why you must listen to the "Voice of the Customer" with caution... Why ninety percent of all opportunity lies hidden beneath the surface... Don't compete, differentiate...   How to use Darwinian divergence to your advantage... Where to start your search for value... The eight values every customer wants and how to use them at the beginning of your search

Chapter 4 – Stop Waste Before It Starts 60

What is Waste? The importance of "fire prevention" versus "fire fighting"... Why everything is the sum of individual tasks or "Ings"... The seven types of solutions that create eighty percent — or more — of high cost, poor quality and slow time to market

Part II – How to Find Your Optimum Solutions 75

Chapter 5 – Search Outside Your Shell 77

The three ways of thinking you must use to have breakthrough insights... When you change the way you look at the world, the world you look at can change... How you can extract new ideas from any experience... Nature has already solved many of our problems... Leonardo's Rule... Start your search for new ideas there... Insights happen when you explore new territories... How to go off your beaten paths to find creative insights

Chapter 6 – Connect Things in a New Way 91

How to go beyond traditional "brainstorming" to focus your idea search for effectiveness and efficiency... The Rule of "Can We?"... Since we tend to use the same problem-solving approaches repeatedly, we usually come up with the same answers... How to find new solutions by looking at your opportunity in many different ways, yet without losing strategic focus... New solutions emerge re-combining both old and new ideas... How to systematically combine different ideas to form new solutions

Chapter 7 – Measure to Learn 110

Why you must measure to learn, not to "prove"... How to build your measurement system... How you can measure what is meaningful, not just what is easy to measure... Make sure you are headed in the right direction first; think about precision later... Avoid sub-optimization... Measure both your *Ilities* and *Ings*... Countering measurement excuses

Part III – The Rest of the Story..127

Chapter 8 – Henry Ford Stumbles..129
Model T dominates auto industry... How Henry Ford broke his own rules and nearly lost an empire... How General Motors recognized the newly ripening bananas

Chapter 9 – The InnovationCUBE©...133
What is the InnovationCUBE©? How you can use the CUBE as your "innovation template" for projects such as cost reduction, value improvement, trade-off decision-making and much more

Appendix I – A Guide to Current Innovation Terminology.......137
Innovation terms you should know as you move ahead in your journey

Appendix II – Further Reading for Innovation Leaders 145
An annotated list of more than 100 books and articles you can use to broaden your understanding and skill

About the Author 155

Acknowledgements

It is impossible to thank all of the individuals and more than 1,000 companies that I have had the pleasure of working with for the past twenty-five years and who contributed to the Rules in this book. Just a few of the companies include General Electric, Microsoft, Zodiac ESCO, Raytheon, NCR, Siemens, Eaton Corporation, Johnson & Johnson, General Dynamics, Briggs & Stratton, Motorola, SKF, Masco, Ford Motor Company, US Navy, Weyerhaeuser, and many, many more.

A special thanks to my wife, Nina, for her common sense and guidance. Additional thanks go to Emily McCreary, Andy Rynberg, John Keogh, Bart Huthwaite, Jr. and all the Institute for Lean Innovation associates, clients and friends who have helped shape the Rules in this book.

Preface

Innovation leadership is the ultimate personal skill. Knowing how to apply it well will give you a clear advantage. The purpose of this book is to give you a step-by-step system for doing just that. It will enable you to change innovation from a mysterious process to a repeatable method. Think of what you read here not only as your personal guide, but also as a model you can use in any organizational setting.

Winding through this book is the story of Henry Ford and the Model T automobile. The narrative device I use is to look at Henry Ford, the innovation giant of his time, to teach us lessons for the future. Why look to the past to prepare for the future? James Burke in his book on the history of innovation, *Connections,* answers this question bluntly: "Because there is nowhere else to look." The past contains clues to the future. The story of Henry Ford, the creation of the innovative Model T, and the near-disaster that befell Ford Motor when Ford broke his own rules of innovation holds plenty of clues. Although the core events I discuss actually happened, what you will read here is my *fictionalized account* of those events.

What Henry Ford was struggling with was that while innovation is difficult to do, it is even *more* difficult to manage. It requires a leadership process. Ford and his team surely would have agreed with Peter F. Drucker, often called the father of modern business management. More than twenty years ago, Drucker wrote in his book, *Innovation and Entrepreneurship,* that innovation is "neither a science nor an art but a practice," capable of being organized as systematic work. In his book, Drucker writes about the "what, when and why" of innovation but, he admits, leaves for others the task of "how to."

Framework for Growth

This book starts where Drucker left off. It will give you the "how to" for making innovation *a systematic process.* This book is not theory. It is based on what I have learned "hands on" from thousands of innovation teams over a span of twenty-five years. The lessons here are also rooted in the history of innovation through the ages. By "systematic work" I don't mean your company's "tool box," "phase gate" process, or portfolio of new products or strategies. I mean what Drucker called for when he used the word "practice" — a systematic, repeatable and measurable process capable of being done as daily work. This book will give you a visible, understandable process you can use to lead a team, a project or your entire enterprise.

Learn and Use in Less Than a Day

No "culture change" is needed for getting you started down the road of innovation excellence. You can use what you learn here to rough out your innovation strategy in *less than a day.* The source of the rules and tools in this book is the InnovationCUBE©, a digital tool for leaders and teams, described in Chapter 9. The secrets of the CUBE have been used for thousands of years by great thinkers, inventors, and corporate leaders, dating all the way back to the Greeks. The CUBE converts their thinking into an innovation leadership process you can use for any project.

The best of success to you on your journey to innovation excellence.

Bart Huthwaite, Sr.
Mackinac Island, Michigan
June, 2007

Introduction

Y our first leadership challenge is to clearly explain the definition of "innovation." Don't rush for Webster's or Wikipedia. They may confuse rather can clarify. Innovation is certainly about new ideas. However, new ideas, as Henry Ford once said, are as "common as ticks on a hound's back." Ideas are good for conversation, but that's about all. They must be converted into practical solutions. What about inventions? Inventions are not innovations. Only a very small percentage of inventions ever succeed in the marketplace. The U.S. Patent Office digital files are overloaded with solutions still looking for customers. Patents are not a good measure of corporate innovation capability.

Innovation is *the process of creating new value with a minimum of waste.* By "new value" I mean things that will benefit both your customer and your enterprise. By "minimum waste" I mean intentionally delivering that value with the absolute minimum of cost and quality loss. I don't mean the "fire fighting" of Lean and Six Sigma on the factory floor. What I do mean is the

1

"fire prevention" of lower cost and better quality created through smarter, innovative thinking.

A Natural Talent?

I have met many folks who can generate a lot of creative ideas. I have met fewer folks who can shape these into unique inventions. However, the task of converting these ideas and inventions into successful innovations, with minimum cost and the best quality, is the real challenge today. During my twenty-five years working with innovation teams I have encountered some people who have this natural talent. You may be one of them. But if you are, you are a very rare bird.

Should you have this natural talent, it can work to your disadvantage. Your innate skill can make a job look so easy that others fail to understand how you do it. Another problem with such talent is that you can find it nearly impossible to teach those who would learn from you. How can you teach what you know instinctively and never had to learn? What you need is the "how to" both for you and others to convert the abstraction called "innovation" into a step-by-step way of thinking and working. The intent of this book is to do just that.

Lessons of the Model T

I have chosen to steer clear from using examples of the present to explain the Rules described here. The reason? Not enough time has passed to rub away the surface success — or failure — to expose the root causes below. (Indeed, many of today's successes may eventually become tomorrow's failures.) The story of Henry Ford and the Model T will illustrate many critical points. The first Model T was shipped in October 1908, and the last came off the line in May 1927. Over that span of time, the root causes of Ford's innovation success, and eventual failure, became very clear.

Almost 100 years have passed since the creation of this remarkable automobile. The modern day corporate practices of lean enterprise, six sigma quality and many others all have roots in the design of the Model T and its production processes. Very importantly, the Model T triggered entirely new ways of thinking about *transactional processes* — how we conduct our day-to-day business practices.

The story begins on a winter Detroit day in early January, 1907. That is when Henry Ford and his team crystallized the ideas that would lead to the launch of the Model T automobile over a year and a half later. Henry Ford was in a fix. A perfectionist by nature, Ford had spent more than a year trying to bring together in his mind what he called the "automobile for the common man." He was going nowhere, so he finally turned to Norval A. Hawkins, whom he called his "process person," for help.

Ford realized that he needed far more than just an automobile. He had to have an end-to-end system that integrated his dealers, plant, suppliers and design department in one seamless system. Hawkins had straightened out the Ford Motor accounting procedures a few months before and would later create the Model T worldwide marketing process that would sell more than fifteen million "Tin Lizzies" by 1927. It was Hawkins who came up with the process that enabled Ford and his team to realize their vision.

Henry Ford with his Model T and Norval Hawkins' innovation cube
visible under his left arm.

I must say at this point that while Norval Hawkins was a real person, his *Rules of Innovation* notebook with its process steps and sketches, and the "innovation cube" tool are my own invention based on the historical data and what I know of both Hawkins and Ford. I do believe, however, that Hawkins would smile and nod his head in agreement with everything in this book.

Ford and his team worked in the cramped "Experimental Room" at the Ford Motor Piquette Avenue plant, tucked away on the building's third floor, away from prying eyes. What went on in that room during those early days has always been a mystery. It is rumored that Henry Ford used a method given to him by his marketing genius, Norval Hawkins, so let's assume that to be true. Let's further imagine that Hawkins also gave Ford what he called an "innovation cube." Ford disliked blueprints and papers. He always asked his pattern makers to convert an idea into a wooden model. Hawkins knew this and created a four-inch cube to illustrate the process that Ford and Hawkins used during those momentous days to launch history's most famous vehicle — the Model T. (More on that cube later.)

Norval A. Hawkins

Henry Ford needed a person like Norval Hawkins. Ford disliked financial accounting as much as he did financiers. He relied on year-end bank statements to let him know whether or not the company was making money. Norval Hawkins first arrived on the Ford scene around 1904. James Couzens hired him as a consultant to straighten out the Ford Company accounting system. Hiring him, in itself, was of note. Hawkins had been convicted on embezzlement charges in 1894, having taken $8,000 as a cashier at the Standard Oil Company. But Hawkins was not your typical ex-con. He was colorful, charming and incisive. On the day he emerged from prison, a crowd of old friends was waiting on the street to cheer him on to better times. Hawkins immediately built a successful consulting and accounting practice in Detroit.

The notes and sketches in Hawkins' *Rules of Innovation* black book, described in the following pages, show that by January 1907 Hawkins had moved from improving Ford's accounting system to helping Ford straighten out his Model T development process. Hawkins was "process" minded. He laid out a step-by-step process that enabled Henry Ford and his Piquette Avenue plant team to build a strategy for the Model T.

He was so successful that by autumn of 1907, Hawkins was hired full time as the Ford sales and commercial manager. Once again, he shifted his process mind to the task of selling Model T autos. Hawkins crafted a selling process that enabled Ford Motor to sell all the autos it could produce. Hawkins stayed with Ford Motor until 1919, and then like many of the original Model T team members, quietly left to join General Motors.

Ralph Waldo Emerson wisely wrote, "Every institution is the lengthened shadow of one man." You will learn how Henry Ford applied the secrets of Norval Hawkins' innovation cube — and then abandoned them, one by one, bringing his company to the brink of disaster.

A Short "Rules of Innovation" Glossary

Hawkins believed that using well worn words can trap us into old ruts of thinking, while new words force our minds to think along different paths. The following list is from his *Rules of Innovation* notebook.

Ilities. High-level attributes of value. The use of *Ility* words enables you to abstract your thinking to a higher, wider plane of thought. Examples: marketability, manufacturability, affordability and maintainability.

Ings. Steps that must be done to deliver an *Ility* value. Each *Ing* takes time, hence costs money and, if done poorly, can result in a quality flaw. Fewer are always better. Examples: designing, testing, manufacturing, installing, maintaining, and more.

Sweet Spot. An innovation "bull's eye." Point at which human *Insight* makes the correct connection between what the market wants and the correct solution for it.

Insight. The human ability to simultaneously combine *Hindsight* (experience) and *Foresight* (imagination) with *Outsight*, the ability to find new knowledge, to create a completely new synthesis of thought.

Outsight. The ability to extract new ideas from experiences outside one's normal range of knowledge. All successful innovations result from a high level of *Outsight*.

The Four Domains. The four life cycle sectors of any offering. These include: Customer Domain, Design Domain, Supply Domain, and Operations Domain. Stakeholders from all four must be part of an innovation effort right at the start. A successful innovation must consider all of these simultaneously.

The Three Sharks. The three forces of constant change: Marketplace, Technology, and Competition. These sharks create both threats and opportunities.

Thing. A solution to a *Want*. *Things* can include products, processes, services, policies and more.

Want. A spoken, or unspoken, value that is desirable. Needs are a subset of *Wants*. Most of the *Things* we want are not "needs."

Waste. A collective word for poor quality, high cost, long time. The creation of all *Waste* starts in the innovation concept stage. Seven types of solutions universally drive eighty percent of all downstream *Waste*.

Zeitgeist. Historical period when simultaneous bursts of innovative thinking occur across many fields. These periods are typically triggered by worldwide change.

What You Will Learn in *The Rules of Innovation*

Chapter 1. Recognize Ripening Bananas

Fatal Flaw: We fail to see opportunities before they are fully ripe.

Timing is everything. This is true whether you are trying to create a technology breakthrough or upgrading an existing product or service to make it more competitive. You do not create opportunities; instead, you must recognize and satisfy them. Henry Ford and his Model T did not create a demand for low cost, family transportation. The Model T met an emerging want. The creators of YouTube met an emerging want. These entrepreneurs saw opportunities before they fully ripened, when they were still green. They were then first in line to pick them when they were ripe. This chapter will show you how to find green bananas.

Chapter 2. Think Backwards

Fatal Flaw: You use conventional thinking.

The innovation process requires you to visualize your goal, no matter how fuzzy that may be. It calls for the ability to see the end from the beginning. Then you must climb your mountain in reverse. This is the opposite of how we normally work in the corporate world. You must learn to think *from* the future *to* the present. It is "backwards thinking." Or, as a savvy engineer from General Electric once described the process to me, "You can't get there from here, but you can get from there to here." This chapter will show you how to do this.

Chapter 3. Discover Hidden Wants

Fatal Flaw: You fail to find new "value gaps" to understand and exploit.

Be very cautious when you listen to the voice of your customer. You will not hear the real "truth." Present day markets do not speak well about what they will want in the future. Homemakers were not shouting out for a microwave oven that could cook a dinner in fifteen minutes. It was just beyond their comprehension. Similarly, customers are limited by their vision of the present. This chapter will show you how to systematically discover opportunities long before your competitors do.

Chapter 4. Stop Waste Before It Starts

Fatal Flaw: You fail to understand and attack the creators of "hidden" cost and poor quality.

I am always amazed how the incandescent light bulb is used as the symbol of innovation. It is actually one of the most inefficient devices known to man and should be the poster child for *Waste*. Just touch an incandescent light bulb and see why. Over ninety percent of the energy used is expended in heat, not light. Remember the definition of a real innovation: "Something that meets an unmet need with high quality and low cost." In this chapter you will learn how to identify and defeat the creators of high cost and poor quality before you ever go to market.

Chapter 5. Search Outside Your Shell

Fatal Flaw: You fail to get out of the ruts of your experience.

New ideas seldom arrive in old directions. The digital watch did not come from established watch companies. The calculator didn't come from the slide rule or adding machine companies. The Internet browser didn't come from Microsoft. This chapter will give you a way to systematically mine new territory for new ideas. It will show you how to break apart any product, service or business practice to find ideas you can borrow, adapt and shape into groundbreaking innovations. The most fertile grounds, you will learn, lie outside your normal experience.

Chapter 6. Connect Things in a New Way

Fatal Flaw: You have no systematic way to combine new ideas into new solutions.

Breakthrough innovations come when you combine existing — even old — things in a new way. Like the creative child with a pail of Legos, you and your team must constantly combine and recombine ideas into different combinations. You must build many "solution sets" to eventually find the best one. You can do this without the expense of prototypes and testing. This chapter shows you how.

Chapter 7. Measure to Learn

Fatal Flaw: You don't apply measurement in a positive way.

That which gets measured gets understood. Innovation measurement must be all about learning. Yet most corporate metrics are really history lessons. They give us information after the "horse is out of the barn." This chapter shows you how to use "real time" measurement to your best advantage.

Chapter 8. Henry Ford Stumbles

Fatal Flaw: You don't link these Rules into an actionable, repeatable process.

Every innovation begins to degrade the day it is launched. This chapter will show you how to link all of the Rules into a seamless innovation process. Henry Ford applied the secrets of the InnovationCUBE© to build his automobile empire. He then broke them one-by-one. This chapter is a sobering account of how success can carry within it the seeds of its own destruction.

Chapter 9. Using the InnovationCUBE©

The InnovationCUBE© integrates the Rules of Innovation into a step-by-step, easy to use process. The CUBE is available on a USB "stick" or installed on an intranet. It is being used today to reduce cost, improve transactional processes, solve quality problems, improve management strategies, and for many other purposes where a higher level of innovation is needed.

Appendix I. A Guide to Current Innovation Terminology

You can use this as a summary of what innovation thought leaders and practitioners are using today. And you can be certain about one thing — this list will already be out of date by the time you read this book.

Appendix II. Further Reading for Innovation Leaders

This is an annotated listing of more than 100 books you may wish to read on your journey to corporate innovation leadership. I have never read a book on innovation, creativity, the history of ideas, inventions, or similar that I have not enjoyed. Some are heavy lifting, while others can be squeezed into a flight from New York to Chicago. This list will enable you to tell one from the other.

PART I

How to Find Your Best Opportunities

CHAPTER ONE

Recognize Ripening Bananas

Fatal Flaw: We fail to see opportunities before they are fully ripe.

The best competition is no competition... Why you must detect opportunities and threats while they are still green... Growth opportunities are not created; they are discovered and exploited... All new Things, or solutions, come from recombining existing knowledge in new ways... The three forces of change keep the doorway of opportunity constantly open... How to recognize these "sharks" and leverage them to your advantage.

Invention is not the mother of necessity. You do not create growth opportunities. Instead, you must discover and exploit them. You must spot new opportunities while they are still "green bananas." These are markets that have yet to ripen. By spotting them first, you will have an advantage. Innovation as the ultimate business weapon begins with the skill of seeing green bananas before they begin to turn yellow.

Seek Green Bananas

Just because your new idea is possible doesn't mean it is practical. And just because it is practical doesn't mean people will

immediately adopt it. There is a time and place for everything. Edison's electric light bulb got a lot of publicity, but gas lights in homes were still being used twenty years later. The portable electric motor was perfected in the late 1890s but really didn't hit mainstream America until the 1920s. Television took twenty-five years to catch on. The Internet really never took off for almost ten years.

Every new innovation is tested by this rule. Green bananas are opportunities that the marketplace does not even know it will want. They are ahead of their time. Green bananas are ideas and solutions that people can't quite yet digest.

Intersection of Three Paths of Knowledge

Finding green bananas and knowing when to harvest them need not be random, mysterious or accidental. They don't just happen "out of thin air." With training, you can see them coming. You can also help ripen and harvest them. An *Innovation Sweet Spot* occurs when you connect three paths of knowledge.

Return with me now to that winter day in the Ford Piquette Avenue automobile plant. You are in a small, hastily built room smelling of new pine and machine oil in the back of the plant's third floor. Ford calls it the "Experimental Room." Dominating one wall is a huge blackboard. Standing before it is Norval Hawkins, Ford's "process person." Scattered around the room are the men who will shape the design of the Model T — and the future of Ford Motor. Rocking back and forth in his mother's "lucky chair" in one corner is Henry Ford, intent on what Hawkins has to say. Chalked on the board behind Hawkins is the sketch shown below. He is explaining to the group how the process of innovation "works."

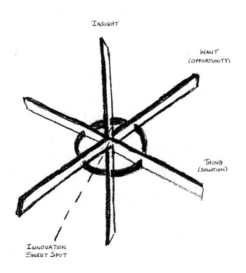

From Norval Hawkins' "Rules of Innovation" notebook, 1907

- **Real Want.** A spoken, or unspoken, value that is desirable. Needs are a subset of *Wants*. Most of the *Things* we want are not "needs."
- **New Thing.** A solution to a *Want*. *Things* can include products, processes, services, policies and more.
- **Insight.** The human ability to simultaneously combine *Hindsight* (experience) and *Foresight* (imagination) with *Outsight* (new experiences) to create a completely new synthesis of thought.

The purpose of Hawkins' sketch is to explain how successful innovations, or "sweet spots," happen. Think of three lines intersecting in a three dimensional space. One is the knowledge of a *Want*. Innovative companies see and exploit *Wants*; they do not "create" them. *Wants* may be spoken or unspoken. The best are the unspoken ones that you can see early and exploit. A second path of knowledge is the knowledge of a solution, or what I call a *Thing*. A *Thing* may be something physical, like a product, or it can be a process, a system or a business strategy.

Some innovations can start with a *Thing* looking for a *Want*. Corporate attics are stuffed with *Things* that have never really effectively gone looking for a *Want*. However, it is the third path of knowledge that must be in place to develop both *Wants* and *Things* to create the intersection called an innovation. This third path is *Insight*. This is the unique human characteristic of "seeing" an untapped *Want* or an underdeveloped *Thing* (solution) and connecting them to creating something new. This is sometimes called the Eureka Moment, the spark that ignites all innovation successes.

A successful innovation comes when you use your skill of *Insight* to find and connect a real *Want* with a real *Thing*. Just having the *Insight* to create a new *Thing*, or technology, is only part of the equation. You must have a real *Want*, that is, a *Want* that is ripe. Think of this ripeness as being a "gap" in the mind of your consumer. It may be latent or just emerging. However, without such a gap, your new idea has little chance of success. You are trying to put a square peg into a non-existent hole. You have a green banana.

Green bananas try to force their way into our minds when we are not quite ready to accept them. Take the example of microwave cooking. This technology was well developed by the late 1950s. But homemakers were not ready to accept the radical idea of bombarding food with World War II radar technology.

Only when homemakers began to see microwave cooking not as a technology, but as a way to reduce time in the kitchen did the microwave oven market begin to accelerate. This acceleration enabled low cost production, which in turn drove down costs to make microwave ovens a part of everyone's kitchen. This same story is true of early television, personal computers and a myriad of other innovations. Green banana innovations are solving *Wants* that have yet to mature.

Some innovations are so green and seem so far ahead of their time that even their contemporaries are unable to understand them. Consider the case of Leonardo DaVinci. We stand in amazement of Leonardo's ability to sketch helicopters, submarines and parachutes. Yet there wasn't even the hint at his time that there was a *Want* for these ideas. What was even more distant was that the *Things*, or technologies, for converting his ideas to reality would only come after hundreds of years of technology advancement.

Trying to digest a green banana before it is ripe can create a bad case of innovation indigestion. Cultivate them, care for them, but don't harvest them until the time is ripe.

Avoid Brown Bananas

Brown bananas are at the opposite end of the ripening scale. Brown bananas are ideas trying to squeeze themselves into an already crowded market space. Their chances for success are slim. Their time is past. Many of these ideas just rot on the tree.

It is true, as President Abraham Lincoln once noted, that "the patent system adds the fuel of interest to the fire of genius."

But there is no guarantee that the genius sparked will result in a bonfire of financial success. Take mousetraps, for example. Who can't forget the old adage, "Build a better mousetrap and the world will beat a path to your door!" Nothing could be further from the truth. Creating new products for market needs already being met seldom pays. The U.S. Patent office has issued more than 2,400 patents for mousetraps. But only a few mousetraps have ever made money. And still dominating the market is the Victor mousetrap, the venerable spring trap wooden device with the big red "V" on it, invented in 1898. The Victor sells for about fifty cents, But it is reported to dominate the mousetrap market with more than a ninety percent share. Most consumers seem satisfied with its performance. However, that doesn't stop new inventors. Every year new mousetrap inventions are submitted.

The moral of this story is to remember that "perfection seldom pays when 'good enough' will do." Your innovation is worthless if it does not deliver an unmet *Want*. And it is wasteful if it exceeds what the market wants.

Ripe Yellow Bananas

Ripe yellow bananas are the ideas that are ready for digestion. These new ideas don't even need to use a new, untried technology. And they can simply be filling an undetected *Want*.

All bananas in a bunch do not ripen at the same time. Technology does not develop in a straight line. Innovations happen in jerks, stops and starts. *Things* and *Wants* must both be mature, or near to maturity.

Bananas ripen faster these days. For centuries, inventions have typically taken twenty to thirty years to go from the first patent or prototype to a widely used product. It took the UPC bar code twenty years before it was finally adopted in 1973. The microwave was developed for cooking shortly after World War II but it took another twenty years before consumers began to buy it in 1967.

The opposite is the story of YouTube, *Time* Magazine's Invention of the Year. More than 70,000 home videos are now uploaded to YouTube every day by people all over the world. The idea of YouTube was a perfectly ripe banana. This phenomenal website went from an idea hatched by three twenty-something guys at a Silicon Valley dinner party to become an immediate market place hit. Why so?

Time says the founders "stumbled" onto the intersection of three revolutions. I like to say they struck their gold at the Sweet Spot of *Wants*, *Things* and *Insight*.

The *Want* is our desire to create and share information together, much like MySpace and Wikipedia. Many consumers are impatient with mainstream media.

The *Things* is the revolution in cheap camcorders and easy-to-use video software. Video production is now so simple and costs so little that it's within the reach of the mass market. The YouTube guys had the *Insight* to connect a *Want* with a *Thing* to reap a $1.6 billion dollar sellout to Google in a little more than a year after its launch. That's a lot of bananas. YouTube had very low barriers standing in the way of its use. Its time was truly ripe.

Your challenge as an innovation leader, at the time I am writing this book, is much like a green banana, perhaps only tinged with a few flecks of yellow. The idea of innovation as a systematic, organized, measurable process that can be universally applied to any product, service or business process is only beginning to ripen. Most corporate innovation initiatives are piecemeal compilations of idea collection, portfolio management, techniques for "open source innovation" and other high level tactics. Your opportunity is to take innovation to the core of your enterprise — to make it a management process that is doable, repeatable and sustainable.

So how do you recognize ripening bananas? The following Rule of *Wants* helps you do just that.

The Rule of *Wants*

> No new idea springs full-blown from a void.
> New innovations emerge from conditions in which
> old ideas no longer work.
>
> — Hawkins' notebook

The most profitable *Wants* are those yet unspoken. These are the *Wants* that have yet to enter the daily mind of the consumer—or your competition. Those who tap into an unspoken *Want* first, and successfully defend their new territory, will be the winners. And that's what Henry Ford did with his Model T for the masses. The Model T tapped into *Wants* that the average American could not even imagine being able to afford. His competition was also caught sleeping. In 1907 the demand for higher priced automobiles was increasing. According to his competitors, Ford was "way off track."

At first "horseless carriages" had a bad reputation. Farmers hated them. Automobiles frightened their horses. Known as "child killers," automobiles careened down streets with ill experienced drivers at the wheel, with brakes that often failed, and with poor tiller-type steering.

Ford "re-purposed" his automobile to meet an untapped want. Ford was a farmer in his youth and he hated farming. Ford always maintained the Model T was designed to help the farmer to get his products to market and ease the tedium of his life. Ford also developed the first mass produced, practical tractor. The Fordson tractor took the technology of the auto and hooked it to a plow.

Timing is Everything

Timing is everything. The market has to be ready, the technology affordable, and the competition sleeping. There is a time for every idea, and when the time comes, the idea springs into the minds of several people simultaneously. The "ripeness" of a culture for a new solution is reflected in the recurrent phenomenon of multiple discovery.

When bananas start to ripen, there are many ways to harvest them. Take for example the case of Wil Durant, the founder of General Motors. While Henry Ford was assembling his first

Model T in 1908, Durant was assembling a number of different fledgling auto companies and heading into a different direction.

For Durant, building General Motors was simply an extension of his earlier career in horse-drawn vehicles. The Durant-Dort Carriage Company, with its coast-to-coast business and its components and assembly plants across the country, was the General Motors of the buggy industry. Durant was applying a successful solution to a new industry by establishing a family of closely related businesses in the motor car industry.

But that concept was still a green banana, requiring more maturing before it could be done well.

The Durant concept and that of Henry Ford were diametrically opposed. Ford "re- purposed" the automobile to meet an untapped *Want*. Ford was completely dedicated to the concept of building a single car capable of meeting basic *Wants*. Ford's idea was that high volume could reduce production costs and, ultimately, the price of the car. This single idea drove his creation of the Model T.

Durant believed that the future would belong to the organization that produced a line of automobiles ranging from light, inexpensive ones which he called Chevrolet (after a famous race driver of his time), to large, expensive automobiles like Oldsmobile (named after Ransom Olds), Buick and Cadillac.

Durant was right but, unfortunately, too far ahead of his time. His concept was still a green banana, requiring more maturing before it could be done well. Durant eventually lost control of General Motors, which struggled until the early 1920s to really find its way.

However, Ford's single car strategy began to fail in the early 1920s. General Motors, now under the direction of Alfred E. Sloan, picked up where Durant left off. As automobile buyers moved into their second or third cars, they began to want more than utility. The market was mature enough for a range of cars, as Sloan said, "built for every purse." Ford's Model T banana was turning brown and his company never again dominated the industry.

But *Wants* are only a part of as successful innovation. You must also be delivering the right *Things* to meet those *Wants*. That is the reason for *The Rule of Things* that I will turn to now.

The Rule of *Things*

Thing. A solution to a *Want*. *Things* can include products, processes, services, policies and more.

Please keep in mind that *Things* may be more than just physical objects. We don't want a 3/8" drill bit. We want a 3/8" hole. Physical objects can only meet part of a *Want*. Also needed are a wide range of tasks throughout a product's life cycle. These include designing, manufacturing, marketing, servicing, learning… the number of tasks—or what I call ings—is endless.

By technology I mean the processes — such as information, labor, materials and money — that convert the *Thing* into products, services, or processes with a higher value. Technology goes well beyond that associated with engineering and manufacturing.

You need not change the physical object much to create a new innovation. Starbucks took what was perceived as a commodity, then added other *Things* such as diversity, location, and ambiance to it to become a worldwide leader.

An innovation is a valued leap from the viewpoint of the consumers. Technology may have nothing to do with its success.

It can come from your ability to find non-obvious opportunity in what, after the fact, *seems* obvious and much wanted by everyone. It can be as simple as noticing what others have overlooked.

Take for example the incandescent light bulb. Roy H. Williams, in his book *The Wizard of Ads,* tells the story of Swan and Sawyer, two brilliant inventors working on opposite sides of the Atlantic. Each of them had plans to produce electric light by running a current through a filament in a vacuum. But it is Thomas Edison who first announced the discovery and who received all the fame and fortune. It doesn't matter that both Swan and Sawyer were further advanced in their experiments, or that it was more than a year later that Edison publicly produced electric light. Edison had the courage to claim he could do it. Sawyer and Swan did not.

Leverage the Three Sharks

How can you recognize these green bananas? One way is to understand and apply the Rule of the Three Sharks. Most times the future is under our very noses. You don't have to be a palm reader to see it, nor do you need a crystal ball. What you do need is a method for *seeing what will destroy the present.* Whether we want to believe it or not, all products and services begin to competitively degrade the very day they are created. By understanding these forces of destruction, we can anticipate them, plan for them and, best of all, shape them to our advantage. The good news is that while these forces are attacking you, they are also attacking your competition. By recognizing and reacting to them sooner, you can achieve advantages, as well as gain footholds in promising new markets. My purpose here is to show you, as a front line innovation leader, how to do this.

These destructive forces are three in number. I call them the Three Sharks of Change.

The first is the Shark of the Marketplace. The second is the Shark of Technology. The third is the most vicious of them all, the Shark of Competition.

By understanding where these sharks are circling now, you can begin to systematically project where and when they will attack in the future. You don't have to lock yourself in an underwater cage. You can swim with these sharks, learn their ways and use them to your benefit.

The Shark of the Marketplace

The Marketplace Shark is driven by ever-changing human *Wants*. Recall that *Wants* are not "needs." Needs are basic. *Wants* are driven by shifting tastes and changing status symbols. No market ever stands still. Consumers are always demanding something that is "better." The trick is to anticipate and respond to these changes before your competitor does. New products and services are continually challenging the dominance of mainstay offerings. For example, the Internet forced Microsoft to entirely revamp its product strategy. Microsoft saw the Internet coming but reacted at only the last minute. And then there are the examples of YouTube, Google and other new products and services that branch out to create new categories of growth, while larger companies look on.

The Shark of Technology

The Technology Shark is always on the move. It can quickly devour you. Remember that when I use the word "technology" I mean the processes that convert existing *Things*, such as information, labor, materials, and money, into new *Things* of higher value to meet existing or new *Wants*. New technology goes well beyond *Things* associated with engineering and manufacturing to include services and even business policies. Take for example the story of the leading manufacturer of slide rules, as described in Steven P. Schnaar's book *Megamistakes*. The company, Keuffel & Esser, was commissioned in 1967 to study the future. Its study produced many interesting findings, some of which came to fruition, but most of which did not. Astonishingly, one of the things it failed to foresee was that within five years the company's own slide rule product would be obsolete, the victim of a substitute product, the electronic calculator. K&E ceased production only a few years later. This perfectly illustrates the fact that technologies go well beyond the engineering and manufacturing of a given product.

The Shark of Competition

The Shark of Competition is the most vicious of all. It can suddenly arrive from many different directions. Please remember that when I use the word "competition," I do *not* mean your existing competitors. Bob Seidensticker, in his book *FutureHype*, reminds us that the most significant new *Things* are likely to come from unexpected sources. Odd as it may seem, the most profitable innovations rarely come from companies that would seem the most likely sources. The digital watch did not come from the established watch companies. Video games didn't come from Parker Brothers or Mattel. The ball-point pen did not come from the fountain-pen industry. The Internet didn't come from Microsoft.

Why did these well-established companies fail to anticipate the attack of the Three Sharks of Change? We may never know the real answers. However, we can guess that none of them used what I call a *Shark Chart,* which is included in our InnovationCUBE© tool. No new project should ever be launched without some prediction about the Three Sharks. The *Shark Chart* predicts, across three time dimensions, where threats and opportunities will come from.

Why use a *Shark Chart* in your job as a front line innovator? Surely your corporate staff has applied all of the latest scientific studies without having you get involved, right? Not necessarily. Common sense is not so common. Your front line team will raise scenarios never imagined by "professional forecasters." And here is a second reason for creating a *Shark Chart*: You must always have your team think in three time dimensions in order to avoid designing yourself into a dead end. One of the best ways to see the future is to use the collective eyes of your front line project team innovators. Always benefit from them.

Don't ever be concerned about a lack of ripening bananas. The Three Sharks assure that green bananas are constantly ripening around us. While these Three Sharks may appear to act independently, there is a rhythm to their speed. This is due in part to an industry's ability to convert or the marketplace's ability to "digest" a new Thing. Just because a new idea is possible doesn't mean people will immediately adopt it. One thing you will want to do want to do with your new idea is to apply the *Rule of the Ripe Banana:* Marketplace *Wants,* technology and the competitive environment have to be in the right alignment. New ideas seldom arrive at precisely the right time.

> The bottom line about bananas: The banana continues to overcome huge hurdles of time and travel. A typical banana travels 4,000 miles from a tropical plantation before being eaten. But, overcoming all these obstacles, they remain near the top of the food hit parade. Americans eat 12 billion bananas a year.

Henry Ford Swam with the Sharks

In the early years of Ford Motor, Henry Ford understood three things that escaped most of his competitors — he knew how to read the Three Sharks. Ford saw a huge market in the vast area and scattered population of the United States. While his competitors saw mid- to high-range cars as the real opportunity, Ford moved ahead with a low-priced vehicle for the mass market.

Ford clearly understood that, while the internal combustion engine had some way to go in reliability compared to steam and electric, that gasoline was plentiful and cheap. Gasoline was also portable, which meant traveling longer distances was possible. Lastly, Ford understood the competition. There were more than 200 auto manufacturers in the United States when he launched the Model T in 1908. He knew the Model T would have to decrease in price every year to keep the competition out of his mass market domain. While competitors were adding features to justify maintaining or raising prices, Ford did the opposite. He stripped the Model T of features, invested in machinery to reduce cost, and drove production higher every month to force costs down even further. He saw his category as being the "utility" auto, with its purpose being dependable, low cost, family transportation.

Norval Hawkins' Ford Model T Shark Chart

Hawkins challenged everyone in the Experimental Room to predict where they saw the Three Sharks in the coming years. He chalked their answers on the blackboard. His purpose: To make sure everyone was thinking ahead and not just about the present. He was setting the "context" for their innovation thinking.

	1908	1910	1912
Marketplace	*High cost prohibits most Americans from even thinking they could ever own an automobile*	*A growing desire for low cost 'whenever I want' transportation, comparable to the advent of the bicycle*	*Cost reduction will bring the automobile within the ability of the wage earner.*
Technology	*Technology of manufacturing gasoline engine to consistent quality standard now possible*	*Technology for manufacturing gasoline engine efficiently will be possible?*	*Manufacturing quantities and technology will reduce costs by more than 75%*
Competitors	*Very little competition in low end of automobile market. Many manufacturers of mid to high range autos.*	*Low end of auto market will become more competitive as market grows.*	*Competition in low end, high volume market will become intense.*

How you can do this: Draw a matrix with the horizontal y-axis divided into three columns. Identify each column as "Step, Stretch and Leap." On the vertical y-axis create three rows, identified as "Marketplace, Technology and Competition." Draw a third line, or z-axis, at an angle where the x and y lines intersect. On this z-axis line use a verb and noun to describe your product or service. Then challenge your team to fill the boxes with bullet statements of possible scenarios. Ask them to use question marks behind these to encourage them to use their imagination. Do this individually first, and then discuss these scenarios as a team.

Summary

- Innovation opportunities are like a bunch of bananas: They are either ripening, ripe or rotten. Early discovery of green bananas gives you time to take action before your competitors.
- Marketplace *Wants* are both spoken and unspoken. Spoken *Wants* are out in the open, with every competitor on the field. Unspoken *Wants* are hidden from view. Recognizing these and quickly taking action in their direction enables you to be the category leader.
- *Things* are solutions for *Wants*. They include new products, services, transactional processes, and policies. They can be technical or non-technical, or a combination of both. Breakthrough *Things* do not normally come from a marketplace leader.
- Three forces of change begin to defeat any innovation from the day it is born. These include the constantly changing *Wants* of the marketplace, improved technology, and new competitive threats. The *Shark Chart* gives you a way to predict emerging *Wants* and *Things*.

CHAPTER TWO

Think Backwards

Fatal Flaw: We limit our success by starting in the wrong direction.

Start with your end-in-view... Why you must climb your mountain in reverse... How to apply "effect-to-cause" thinking... Why identical innovations occur at the same time... Importance of speed-to-market... Why a stakeholder who is not part of the solution will become part of the problem... How to get all stakeholders climbing the same mountain... Why your first ideas are never the best... How to move back and forth between opportunities and solutions to find the best path up your mountain.

Take yourself back again to that January day in Detroit over 100 years ago. You are in a cramped room smelling of machine oil and sweat in the back of the Ford Motor Car Company's Piquette plant. A wiry Henry Ford in his early forties is standing in front of a huge blackboard in the Experimental Room. On the table in front of him lies a bunch of green, unripe bananas. Ford chalks the number $500 on the board for all to see.

Imagine that Ford is addressing the team members in the room: "We have to lower the price of the new Model T to $500 — or less. Remember, gentlemen, that the average wage of mechanics, clerks, small shopkeepers, school teachers, and ministers is between $600 and $900 per year. A $2,500 auto is just too expensive to reach this huge market.

"And let's remember another thing," Ford continued, "maintenance costs of most cars today would make even a Vanderbilt wince. We want to make our parts as reliable and as well supplied — at the lowest possible cost — as a Singer sewing machine. The banana of opportunity is ripening fast. If we build the Model T right, the world will buy it."

History shows that people who have accomplished anything usually did not know exactly *how* they were going to do it. They only knew they *were going to do it*. You must do the same. Innovation leadership requires that you see the end before the beginning. You then must be able to communicate your vision to the people who will work with you to make it a reality. Your vision of your mountaintop might be somewhat fuzzy at first and will certainly change. But don't be too concerned about that. The records show that no company ever ended up at the same destination first aimed for.

Start With Your End-In-View

You first must make sure those you are relying on understand the concept of "reverse thinking." This is much like climbing a mountain backwards. It is the safest way to go to a place where no one has been before. Mt. Everest climbers use this technique all the time. There is no clear path up the mountain, so you must first find one by climbing your mountain mentally in reverse.

You can't get there from here, but you can get from there to here. Your thinking process must move you from the future to the present.

At first, don't be too worried about the "how" of getting there. Focus on creating "pull." Concentrate on creating a compelling vision of where you are going. My good friend, Colonel Larry Stewart, compares this to a rubber band stretched from your thumb. By pulling it *from* the destination *to* where you are now, you are creating tension in the right direction. It brings the objective to you and carries you to your objective.

Why Reverse Thinking Works

This kind of "reverse thinking" is the opposite of how we normally think. We most often use "deductive thinking." Deductive thinking has us starting from the current state and then working forward. We look at the resources presently available to us and ask: What is the best way to use these resources to achieve possible outcomes? But when we work backward, we must first determine the goal that's worthy of our efforts, and then we get to the point of asking: What resources must we currently develop to achieve our goal? When we work forward, we work only within our means. When we work backwards, we may find ourselves stretching our resources and having to work beyond our means, but in so doing we reach heights we never could have achieved through conventional planning.

By working backwards you will find yourself creating ways to do more with less — finding ways to work smarter rather than harder. Working backwards is a way of forcing more of the problems to the surface at the outset rather than at the end — creating chaos at the beginning to help order emerge at the end. Of course, delivering an innovation requires both inductive and deductive thinking. However, inductive thinking must always precede deductive thinking.

How Ford Did It

Henry Ford had the intuitive ability to do this. In his early days, he never lacked for help. He attracted others to his cause like a magnet. In the 1890s Ford had five people working with him on his "quadricycle" at no cost to him. Ford historian Allan Nevins, in his classic book *Ford: The Times, The Man, The Company,* describes Ford's mentality well: "It was his instinct, in the drive toward his dream, to harness anyone who could help pull him in that direction."

Another Ford historian, Douglas Brinkley agrees: "What Henry Ford could do, brilliantly, was lead. It was he who foresaw the growth of the automobile market. He pushed his company incessantly toward higher and higher production to meet this growth. Although he was known for mumbling and stammering through any sort of public speaking engagement, Ford had no trouble communicating his grand plans to his employees at Ford Motor and inspiring them thereby to greater achievements."

Henry Ford recognized that the world was changing rapidly. The mass of Americans would not be satisfied to watch only the rich afford the ability to go where and when they wished. Years earlier, the bicycle had democratized the idea of personal transportation. Ford's idea was to "democratize" the automobile by making it affordable by the average American family. Ford saw a world most of his automobile competitors could not even imagine.

The Spirit of Innovation

Steven Schnaars, in his book *Megamistakes,* uses the German word *Zeitgeist* to describe the phenomenon of innovation beginning to happen simultaneously worldwide. The idea of the *Zeitgeist* holds that there is a characteristic "spirit of the times" marked by social, technical, intellectual and political trends of that era. These are ideas whose time has come. It means that the conditions for germinating new ideas are ripening, like Henry Ford's bunch of green bananas.

In the Middle Ages, a *Zeitgeist* occurred when the Turks in 1470 disrupted the overland trade routes to Europe and the Far East. This shut down the European supply of pepper, the spice that stood between meat-eating Europeans and starvation. No spice except pepper made heavily salted meat edible, and salting was the dominant form of meat preservation. This caused pepper to be in short supply and prices skyrocketed. As a result, European explorers sailed west and south in search of new sea passages to the Orient. The discovery of America by Columbus was a by-product of this search for pepper.

Similarly, in the early 1930s, gathering war clouds sparked a leap in aviation technology. Sir Frank Whittle, the famous British inventor, successfully demonstrated in 1937 the first jet aircraft engine. By coincidence, Hans von Ohain, a German scientist, built a similar jet engine that same year. Only after the end of World War II, when the two inventors finally met one another, did they learn they had reached the same technical solution through separate thought. When asked how two scientists — unknown to each other and unaware of each other's work — could develop the same device, Whittle replied, "The tree of scientific knowledge tends to bear its fruit at the same time."

It was the same with the birth of the automobile industry. It is estimated that, by 1895, more than fifty people were independently working on automobiles across America, most thinking that they were the sole leader in this idea. Why did so many different and widely separated persons have the same thoughts at the same time?

The Bicycle Triggers a *Zeitgiest*

If pepper set in motion the *Zeitgiest* of world discovery, the bicycle did the same for the automobile industry. It is a common misconception that the internal combustion gasoline engine created the burst of energy that led to the automobile phenomenon. Nothing could be further from the truth — it was the bicycle.

Hiram Maxim, an early automobile pioneer, said that the emergence of the low-cost bicycle directly led to the automobile. The bicycle "directed men's minds to the possibilities of independent, long distance travel over ordinary roads." The low cost bicycle also ignited bursts of innovation. All the major automobile innovators, from Daimler to Ford, started with the bicycle as their mass production model. Even Wilbur and Orville Wright were bicycle builders.

Prior to the bicycle, the only means of transportation freedom was the horse.Horse-drawn carriages were expensive, though, and horses were costly to keep and produced several pounds of manure a day.Horses were out of the question for the ordinary city dweller due to space constraints and sanitation issues. Farmers had horses for work and transportation to town, but they also had grazing ground and room for stables.

The bicycle, on the other hand, was a cheap, clean and efficient mode of personal transportation. The bicycle was first popularized in Paris in the 1860s and by the 1880s the bicycle craze was sweeping across America. The mass production of the drive chain "safety bicycle" meant that both men and women were free to go wherever they wanted, whenever they wanted.

For speed, efficiency and cost, a bicycle will beat a horse or a horse-drawn carriage every time. I can state this as an absolute fact. I live on Mackinac Island, Michigan, where all forms of motorized transportation are absolutely banned.In 1901, our City Council passed an ordinance banning motorized vehicles. The instigators of this ordinance were the horse-drawn tour drivers who wanted no competition from the downstate Detroit "horseless carriages." The result is a transportation economy frozen at the turn of the twentieth century. My world is one of bicycles and horse-drawn taxis and drays.

While the owners of the present-day horse-drawn taxi company are our friends, I much prefer riding my bicycle. The reason? I can go wherever I want, when I want, in one-third the time and at no cost. I conduct this experiment of bicycle against horse every day, and the bicycle always wins. The only downside is that the bicycle requires manpower to propel it.

The arrival of the bicycle in late nineteenth-century America created the *Want* of freedom of personal transportation. This, in turn, stimulated the quest for *Things* that could provide this freedom for even greater distances and speed. Thus the automobile began as an extension of the bicycle. Ford's initial idea was to build a two-cylinder engine to propel a bicycle. In Germany, Daimler was also experimenting with a motorized bicycle. In 1891, Ransom Olds built a steam-powered tricycle. A while later, Henry Ford built his gasoline powered Quadricycle.

The French coined the word "automobile" and it stuck. Ford's product became for a time the "Fordmobile," which was short-lived. But the name "Oldsmobile," named after its creator, Ransom Olds, did stick. This points out that a common trait: we are slow to leave the comfort of our old paradigms. We use terms that we know to describe things that we don't know or are not sure of.

Ford's challenge to his team in the Experimental Room was different than that of other auto manufacturers: "The question is not, 'How much can we get for this car?' but 'How low can we sell it and make a small margin on each one? How many cars must be turned out to get the lowest cost per car, and will the demand absorb this tremendous output?'"

Ford focused on all the *Things* that had to be accomplished to mass produce the Model T while reducing it to an affordable price range and thus creating the new category of the family car.

His insight was to recognize that mass production is more than quantity production. It means the integration of design simplification, part standardization, precision machining, coordinated production, continuous motion, and the reduction of labor of all kinds anywhere, including the user of the automobile.

Ford was so confident that he began advertising the (as yet undeveloped) features of the Model T, a full year before the Model T entered into design. "Our idea," he told the press, "is to build a high-grade, practical automobile, one that will do any reasonable service, that can be maintained at a reasonable expense, and at as near $500 as it is possible to make it, thus raising the automobile out of the list of luxuries and bringing it to the point where the average American citizen may own and enjoy his automobile."

Get Your Stakeholders on Board

When a *Zeitgeist* erupts, it offers opportunities across an entire value chain. These opportunities range from customers all the way back to your company and then further back to your supply chain. If you don't invite these "stakeholders" into your creative process, you will miss out on including their ideas. What's worse, a stakeholder who is not part of the solution can turn into part of the problem later. Disgruntled stakeholders can even turn into resistors and saboteurs. You must engage stakeholders in your innovation process right from the beginning. If the goal is meaningful to them, they will become enthusiastic. People only support what they create.

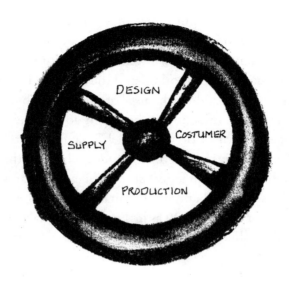

Four Domains Wheel
From Norval Hawkins' "Rules of Innovation" notebook, 1907

Both Ford and Hawkins knew this well. Hawkins used what he called Four Domain thinking. He thought of stakeholders as representing four "domains," each of which was needed to operate in harmony. He called these the Customer Domain, the Design Domain, the Supply Chain Domain, and the Operations Domain. All of these areas would have a major impact on a product's success.

This is the reason for including stakeholders from all four domains, so all visions and motivations can be revealed right from the start. Remember — if a stakeholder is not part of the solution, he or she will become part of the problem later. However, Ford and Hawkins also realized that stakeholders do not have to all participate in a project for the same reason, and shared vision does not have to coincide with shared motivations.

For example, the two Dodge brothers, Horace and John, were part of the Experimental Room team right from the start. They supplied all the engines and drive trains to Ford's specifications. Ford also invited Harvey Firestone in. His Firestone Tire Company would supply all the Model T tires. The team also included Alfred E. Sloan, president of the Hyatt Roller Bearing Company (who later became president of General Motors). Hyatt Roller Bearings were a critical component if the Model T, used to counter the abuse of rough country roads. All these suppliers had separate motives, but all were united in the idea of making the Model T the world's dominant automobile.

Stakeholders are more easily convinced by what they themselves discover than by what you direct them to do. The trick is to get stakeholders meaningfully involved at the front end of any project, and then keep them involved without long, exhausting meetings.

Ford knew that getting the right stakeholders involved at the start of the Model T development was essential. He knew that sometimes the most seemingly insignificant stakeholder can have a major impact on a product's success. While innovation is a highly creative human process that is often individualistic in its source, it requires the organized efforts of many others to be effective. When you include stakeholders from all domains in decision-making, don't be too concerned about having total agreement at the start. Perception of what is "right" is a matter of interpretation, which means that different points of view are going to be part of every interaction.

Iterative Thinking

We rarely get our innovation equation right the first time out of the chute. This was the case with Ford. Between the founding of Ford Motor Company in 1903 and the advent of the Model T, Ford Motor had offered more than a dozen different models, each with a different target market. The *Wants* of the fledgling automobile market were chaotic. Was electric better than gasoline power? Would steam power still be in the running? Was a two passenger vehicle better than a four passenger? Some of the more than 200 companies manufacturing automobiles offered all three power options. However, by 1907, Henry Ford perceived that the *Want* banana was quickly ripening in a new direction. He and his team designed the Model T for the then almost non-existent, but quickly emerging, mass market. Ford defied "logic" by going where the market *would be,* not where it was.

Ford encouraged his people never to think in a straight line. Innovative thought requires "zig-zag"

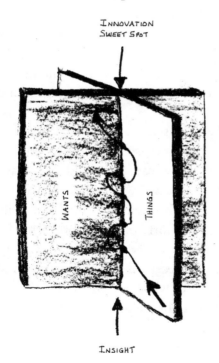

From Norval Hawkins' "Rules of Innovation" notebook, 1907

thinking. Your mind must constantly move between different definitions of the problem (the *Want*) and the solution (the *Thing*) until an *Insight* (a Eureka!) is discovered to connect one to the other. Research suggests that in over ninety percent of all successful new businesses, historically, the strategy that the founders had originally decided to pursue was *not* the strategy that ultimately led to the business's success. Intel, Wal-Mart, and a host of other companies each saw a viable strategy emerge that was substantially different than their founders had envisioned. But once the model was clear, they executed that strategy aggressively.

Such iterative "zig-zag" thinking is contrary to our usual pattern of thought and can seem like scratching your nails across a chalkboard. . We see a problem, usually well defined, and we systematically apply time-tested theories and tools to solve it. So your first task is not so much to come up with new ideas, but to first escape from the clutches of old ones. Innovative thinking starts with breaking the tight bonds of our present frames of thinking.

Messy Business

Problem and solution (*Wants* and *Things*) must co-evolve in the innovation process.

This makes parallel thinking messy business. Nigel Cross, in his classic *Engineering Design Methods: Strategies for Product Design*, explains the dilemma well. He says that even a fairly precise problem statement gives little indication of what a solution must be. Problem and solution must be developed in parallel, sometimes leading to a creative redefinition of the problem, or to a solution that lies outside the boundaries of what was previously assumed to be possible. This makes innovation also a messy business. Your difficulties are twofold: understanding the problem (the *Want*) and finding a solution (the *Thing*). They both occupy a space that is initially undefined. The key to successful innovation is the effective, and efficient, management of this dual exploration of both the "problem space" and the "solution space."

A grandstand view of history shows that connecting even *known Wants* with *known Things* is not so obvious. New *Things* have languished for decades before they were recognized as being valuable. Likewise, new *Wants* emerge and exist for decades before some combination of *Things* is found to satisfy them. The time span between the invention of canned food and the can opener was more than fifty years. What was used during this technology gap? Labels on cans advised using a "hammer and chisel"!

We get ourselves in trouble when we are overly rigid in defining either the *Wants* or the *Things* to satisfy them. We end up in a blind alley at one or another of these dual planes. The *Things* that you find may be something completely different or never imagined possible. Or they can be the discovery of *Wants* you never realized that the market wanted. It is this kind of uncertainly that makes innovation such a challenging activity. The Three Sharks assure that *Wants* and *Things* are both constantly changing. Despite his early success, Henry Ford's critical mistake was failing to change the Model T when the market changed. He tenaciously clung to its "one for all" design while General Motors was moving down the path of "an auto for every pocket and purse."

Surface Boundaries and Hurdles

The *Boundaries and Hurdles Tool* gets you and your stakeholders headed in the right direction. *Boundaries* help set the margins of your ascent. Going beyond them is wasteful and even dangerous. The architect Frank Lloyd Wright always told his students, "Limits are an artist's best friend." Tight limits force us to think beyond conventional solutions and find answers we might not have otherwise discovered. When you are challenged to deliver a new innovation with a small budget and tight deadline, you have probably found that you were more resourceful than if you had been granted a lot of money and time.

Hurdles are challenges that you must overcome to reach your mountaintop. Identifying them early gives you time to solve them, and gets everybody roped together and prepared to head for the correct mountaintop.

Hawkins' Boundaries and Hurdles Blackboard Tool

Goal: First Model T Shipped June 1, 1908

	Boundaries	*Hurdles*
Customer Domain	Five passengers. $850 price	Weak sales network? Can we make money at $850?
Design Domain	Gasoline engine. 20 hp.	Do we have enough time to complete design by early 1908?
Supply Domain	Dodge Bros. drive train Firestone tires Hyatt Bearings	Will Dodge Bros. have the quantity, quality we need?
Operations Domain	Assembly only at Piquette Avenue plant.	Do we have enough skilled labor? Do we have the space required?

Like all the tools you will learn in this book, think of it as a three dimensional matrix. On the horizontal Z axis draw two columns, one headed with the word "Boundaries," and the other with "Hurdles." Boundaries set the margins of your ascent; remember that going beyond them is wasteful and can be dangerous. Hurdles are challenges that you must overcome to reach your mountaintop; remember that identifying them early gives you time to solve them. On the vertical Y axis, list the Four Stakeholder Domains of Customer, Design, Supply Chain and Operations. On the Z axis, describe your objective in a few bullet points, each of which is at a high level. Then ask each of your stakeholders and team members to individually write bullet points in each box. Question marks should be used when a person is uncertain about his or her bullet point. Then integrate all of these thoughts into one summary. You will be amazed at how different each stakeholder perspective varies from one domain to another. Benefits of this tool include:

- Surfacing misalignments with your goal
- Understanding individual fears and concerns so that you can deal with them effectively
- Discovering false boundaries that limit possibilities.

- A supplier who is not part of the solution (design) will be part of the problem (production) later.
- To go to a new destination you must first leave an old one.
- The best way up a mountain is never a straight line.
- If man had never charted a path into the unknown, he would never have discovered the known.

—From Hawkins' notebook

Summary

- Innovation requires "inductive" thinking. You set a goal and then work backward to the present to discover the best way to climb your mountain. This frees you of your current resource constraints.
- Any innovation requires speed. The idea of *Zeitgeist* says that the things that have awakened you to a new opportunity have also done that to others.
- Stakeholders who are not part of defining and solving your innovation challenge will become part of the problem later. Get them on board immediately by using the *Boundaries and Hurdles* tool.
- Iterative "zig-zag" thinking is a hallmark of innovation. You must explain that innovation is "messy business." It requires moving constantly between re-defining the *Wants* and the *Things* to answer them. The route to a profitable solution has many twists and turns.

CHAPTER THREE

Explore for Hidden *Wants*

Fatal Flaw: We Fail to Discover New Values to Deliver

Why you must listen to the "Voice of the Customer" with caution... Why ninety percent of all opportunity lies hidden beneath the surface... Don't compete, differentiate... How to use Darwinian divergence to your advantage... Where to start your search for value... The eight values every customer wants and how to use them at the beginning of your search.

This chapter will show you how to discover hidden *Wants*. The most promising kinds of *Wants* are those that are unspoken. If you have the power of hearing unspoken wants before anyone else, and developing the right *Things* to satisfy them, you will be a winner.

Yellow banana innovation strategies exploit a change that has already occurred. They satisfy a need that already exists. But in green banana strategy, you identify emerging *Wants* and bring them to

the surface. Your aim is exploiting this kind of hidden *Want*. The trick is to have some idea about when the *Want* is going to be "ripe." Is the user going to be receptive, indifferent, or actively resistant?

Hear Unspoken *Wants*

Be cautious when you listen to the voice of the customer. Customers do not tell you the entire truth. They don't intend to hold their *Wants* back; it's just that they don't realize they have them, or are unable to express them in words.

Roy H. Williams, in his book *The Wizard of Ads,* expresses this well: "There is a profound difference between what a customer says she wants and what she truly wants. The diamond buyer says she wants a GIA certificate diamond of G color and VVS clarity when all she really wants is to see people's eyebrows jump to their foreheads as they exclaim, 'Is that real?' Similarly, a customer may say he wants a car wax when all he really wants is a shiny car!

The marketplace seldom speaks the "real" truth about its real *Wants*. The reason? Customers typically speak in the present tense. They describe what they would like today and, most times, don't do a very good job of expressing their real needs. Your challenge is to discover these unspoken *Wants*.

Starting Your Exploration

Your exploration for hidden *Wants* must begin with "value abstraction." Value abstraction is the technique of going to a high enough altitude to see the forest, not just the trees. When you go high enough, you are able to see opportunities you never imagined before. It's much like the Civil War balloonist who saw a pattern of battle evolving and opportunities arising that no one could ever imagine.

Take, for example, Henry Ford and his idea of a "universal car." Most automakers in 1907 saw the market as being those who were able to pay $2,500 or more for a car. They were looking at only five percent of the U.S. population.Ford's thinking went high enough to see the other ninety-five percent of the population who would be able to afford a car if he could get the price below $500. Ford's competitors defined "affordability" as a minimum of $2,500. Ford took a higher view and eventually drove down the price of a Model T to less than $350. It is when you begin at too low an altitude that you get in trouble. You limit your possibilities and you may miss major opportunities.

The Eight Basic Value Questions

There are two main reasons why many value searches flounder. The first is that you may start at too low a level. Specific problem statements can lead to quicker solutions, but less conceptual creativity. You constrain your search by using specifications rather than abstractions. For example, you define "performance" within a narrow band of technical constraints. Such constraints leave you little room for looking at a wide range of alternatives and you are forced to see only the trees, not the entire forest.

The second reason is that you do not have a starting point. Value abstraction must be high enough to encompass all customer *Wants;* however, the abstractions must be clear enough to act as a starting point for finding new value. For example, "customer satisfaction" is an abstraction you most certainly believe in, but the term is at too high a level to act as a beginning point.

Every quest for the right value set must have a starting point. My experience shows that there are eight universal "core values" all customers want. These may be stated in different words or languages; however our experience over the past twenty-five years consistently shows these eight are basic. You can use them as a first step in defining your overarching value search.

Let's return again to Ford's Experimental Room where Norval Hawkins chalked these eight terms on the blackboard and is challenging everyone in the room, including Henry Ford, to define them for the Model T.

Performability

Hawkins' Notes: The Model T must be able to go anywhere and anytime. The goal will be utility. We will build it light, with a lot of road clearance. We will reduce weight to the absolute minimum. This means we can build an auto of twenty horsepower that can run circles around a heavier one with twice the horsepower.

Ask yourself: What defines performance in the context of the customers we are attempting to reach? Are we putting in too much performance?

Affordability

Hawkins' Notes: The Model T will be the lowest priced auto on the market. Mr. Ford wants to drive the selling price down below $500! His goal is to ship 1,000 Model Ts a day!

Ask yourself: Cost is always relative to the value received. However, value is always defined by the customer. Are we delivering too much value, more than the customer really wants for the price?

Featurability

Hawkins' Notes: Mr. Ford wants to strip the Model T to the basics. Should buyers want "extras," they can have them installed after their purchase.

Ask yourself: What features can we eliminate without a major decrease in value? Consider that every feature, whether it is asso-

ciated with a product or a service, takes time, hence costs money, and if done wrong can create a quality flaw (more on this in Chapter 4, "Stop Waste Before It Starts").

Deliverability

Hawkins' Notes: Mr. Ford insists on no special orders. We will set our production for the year and stick to it. Volume, volume, volume, he says. Only that way will we be able to drive the price down to the lowest in the marketplace.

Ask yourself: Can we deliver our offering precisely when it is needed? Can we gain a competitive advantage by doing this? How must we structure our service or product's architecture to make sure we meet the right tempo of our customers' demands?

Useability

Hawkins' Notes: Mr. Ford believes the ideal number of pages in an instruction manual is zero. His challenge: How can we design the Model T so consumers can learn how to drive it and maintain it in minutes?

Ask yourself: What can we do to eliminate the need for learning? How can make our product as easy as possible to use? Can we eliminate the need for skill?

Maintainability

Hawkins' Notes: Mr. Ford wants the Model T to be so simple that it can be serviced by a city person with normal tools. He wants it built light to reduce wear on tires.

Ask yourself: Can we gain a competitive advantage by reducing all skill required for maintenance, or eliminating maintenance entirely?

Durability

Hawkins' Notes: Mr. Ford wants the Model T to be a farmer's auto, able to go across any road or field. He wants it built light with a high clearance. This will make it easier to get out of ruts.

Ask yourself: Is there a competitive advantage to making our product more durable? Is it too durable?

Imageability

Hawkins' Notes: You never get a second chance to make a first impression. Mr. Ford wants the Model T to convey the image of trust and quality. He says we will enter every endurance race in America and Europe to demonstrate this. He has no interest in style.

Ask yourself: What is the fundamental image that our offering now conveys? How can we change this? Should we change it?

Why *Ilities* ?

The use of abstractive words such as "*Ilities*" enables you to apply the principle of *problem restructuring*. For example, the standard procedure in physical science is to make observations or collect systematic data and to derive principles and theories from them. This imposes limitations on the breadth of your innovation thinking. Albert Einstein despaired of creating new knowledge from already existing knowledge. How, he said, can the conclusion go beyond the premises? So he reversed this procedure and worked at a higher level of abstraction. This bold stance enabled him to create theories others could not. Einstein took an abstraction as his starting premise and simply reasoned from the abstractions to a theory. This approach ran counter to what others were unwilling to accept because such abstractions could not be proven by experiments of the day.

You may ask, why use only eight attributes? Too many choices in value abstraction often leads to taking the easy way out. Focus is needed. A good photographer will get much closer to his subject, excluding everything from the picture except what matters most. The principle of focus is to exclude by choice that which matters less, so that we may give our undivided attention to that which matters most. Our experience shows that eight is the maximum number of values that can be juggled at one time.

Ford's Value Abstraction

A Ford designer said, "Mr. Ford wanted everybody to 'see' the values first, and then design the car to meet those values." While other manufacturers made running changes to meet what their salesmen told them their wealthy customers wanted, right from the start the Model T team stepped up higher to get a clearer view. Ford once said, "Our effort must be in the direction of simplicity. People in general have so little and it costs so much to buy even the barest necessities, because nearly everything that we make is much more complex than it needs to be."

Hawkins' Ility Chart Blackboard Tool

Universal Customer Values	Ford Model T Team Definitions
Peformability	"Utility… go anywhere, anytime"
Affordability	"Anybody can afford to buy one… goal less than $500"
Featureability	"Absolute minimum… extras can by added by customer"
Deliverability	"Off the dealer's floor… no special orders"
Useability	"So easy the wife can drive it… minimum training"
Maintainability	"Low skill, simple tools"
Durability	"Built for country roads…can run through mud and snow"
Imageability	"Common man's auto… sensible, not stylish"

Hawkins chalked the eight "universal customer values" on the blackboard. He then challenged the Model T team to come up with definitions for each. He had Henry Ford start the discussion, with each person, in turn, adding his thoughts. When the process was done, the Model T team had a high level, unified view of what the ideal Model T would be. While there would be many arguments about specific requirements, the entire team never questioned this overall view they had originally agreed to.

Summary

- Be cautious when listening to the voice of the customer. Customers have a limited view of what is really possible.
- The best kinds of *Wants* are the unspoken ones. When you can find and deliver them before your competitor, you will gain a distinctive advantage.
- Don't compete, differentiate. Explore for advantages that may take you in an opposite direction than where the market is going.
- All customers want eight basic values. Use these as a starting point and find new definitions for each of them.

CHAPTER FOUR

Stop *Waste* Before It Starts

Fatal Flaw: You fail to find and attack the creators of hidden cost and poor quality.

What is Waste?... The importance of "fire prevention" versus "fire fighting"... Why everything is the sum of individual tasks or "Ings"... The seven types of solutions that create eighty percent — or more — of high cost, poor quality and slow time to market

A *Want* is like a two-sided coin. One is the side of *Value*, the *Ilities* that your customer seeks. The other is the side of *Ings*, the tasks that must be done to deliver that *Want*. Your goal must be to optimize the *Ilities* and minimize the *Ings*. For example, Ford constantly struggled to reduce to an absolute minimum the number of steps required to operate a Model T. He wanted easy "operate-ability." If the Model T was truly to be an auto for the masses, Ford knew he had to minimize the "learn-ings" of what was normally considered to be a high tech machine.

Many times you can get the *Value* side correct, but create flaws by imbedding the wastes of high cost and poor quality. *Wastes*, like *Values*, are set in motion at the early conceptual, or incubation, stage. Most of such wastes are hidden from view. You cannot "see" them in the usual sense of the word since they have not "hardened" into drawings or prototypes. So they are missed, only to emerge into realities on the factory floor or in your cus-tomer's domain. In this chapter you will learn how to "see" and attack such *hidden Waste*.

Be an *Ing* Thinker

Begin by thinking of all *Things* as the sum of their collective *Ings*. These are the tasks, or "jobs to be done," in order to create, sustain, and even dispose of a product, process or service along its life cycle. They occur in all the Four Domains described in Chapter 2. Waste prevention begins with imagining all these *Ings*.Your goal is to deliver value, or *Wants*, with as few *Ings* as possible. Why is this so? As I mentioned in the previous chapter, each of these tasks takes time, hence costs money, but also, if done improperly, can create a quality flaw. A very basic waste reduction technique is simply to reduce process steps. These include "hard" processes such as manufacturing as well as "soft" processes in the transactional areas, such as back office operations and marketing functions.

Think of *Ings* as being the molecules that can be combined to create the *Ilities* you desire. The problem is that many of these

Ings are invisible, like molecules, at the concept stage of your effort. They only become visible downstream where they are combined and used to create and sustain your offering.

Henry Ford was an intuitive *Ing* thinker. He saw cost where others saw nothing. Ford believed that the Model T would be the sum of the tasks required to create it, manufacture it, and sustain it in the customer's domain. Ford Motor consistently beat out its competitors because of his relentless attack on what he called the "unseen creators of waste." These were the evils of sorting, handling, fitting, fine-tuning, inspecting and many hundreds more. Ford's greatest contribution was not the moving assembly line, as most think. It was his ability to see the Model T as a system of tasks, from beginning to end. He would constantly try to integrate these tasks into one continuing process, from growing the trees for his wooden body parts to converting the wood waste into charcoal briquettes so that it could be reused.

Ford himself regarded his greatest contribution to automotive history as the design of the Model T. He saw the production of the Model T flowing naturally from its simplicity of concept and design. Even when the moving assembly line did arrive at Ford Motor in 1914, seven years after the design of the Model T, it was the modularization of the design, elimination of manual precision, standardization of parts, and similar ideas that enabled the moving assembly line, as well as all the other sub-assembly lines feeding it, to become a reality. Other auto manufacturers attempted to apply moving assembly but struggled and failed due to the design of their vehicles.

According to Norval Hawkins' notebook, Henry Ford himself created the idea of "*Ing* thinking." On the Experimental Room blackboard, Ford drew a cluster of circles, both large and small. Some he chalked in solid and then connected those with lines. The others were left floating in the space of the huge board.

"Gentlemen," he said, "the total of what I have drawn here is our present Model N auto. Each of these circles represents a task, or *Ing*, that must be done either in our factory or by our customers. We will be successful when we have eliminated a minimum of fifty percent of these tasks, leaving only the ones I have chalked in and connected to one another. The unconnected circles are waste. We will eliminate them by design!" Ford then grabbed a rag and vigorously erased the unconnected circles. He raised *Ing* thinking to a new level, by taking complexity reduction to the early innovation phase.

Prevent *Waste*

Remember that there is nothing as wasteful as doing something with great efficiency that need not be done at all. We unknowingly create the environment for waste downstream in the factory and in the marketplace. The reason: We have few ways at the incubation stage to illuminate, or predict, what kind of *Waste* will occur. Once imbedded in a product or service, *Waste* is tough to eliminate.

Corporate America has spent billions of dollars on curing manufacturing and customer service problems that could have been avoided in the early innovation stage. Lean manufacturing is focused on eliminating non-value tasks on the factory floor. This includes identifying and reducing such "hidden" process steps as part moving, storing, inspecting, and many more. Manufacturing complexity reduction is the primary target. Six Sigma tools, and the cadres of Black Belts who administer them, are aimed at identifying the root cause drivers of quality failures both on the factory floor and beyond. Product and process variability is the chief enemy.

What is needed today is a stronger dose of *waste prevention.* Many products are designed with value in mind; however, little thought is given to waste prevention. One reason is that the kinds of waste these products create are well off the corporate accounting radar screen. There has been no system for detecting their potential and reducing their effect at the early concept stage. Remember that your concept stage will drive most of your success. Once your concept hardens into a prototype, or solidifies in the minds of your team and management, your flexibility and receptivity for change shrinks rapidly. The pressures of time and money demand that you "get on with it." You enter into a stage of fine-tuning and implementing, rather than innovating.

"Fire prevention always beats fire fighting" is another way of looking at it. I am a great believer in both Six Sigma and Lean. The past years have been a gigantic learning experience. Six Sigma gave us tools and a philosophy for systematically identifying the creators of poor quality and eliminating them. Lean gave us the Toyota-type tools for identifying waste on the factory floor. Both have begun the climb upstream to the real cause of the problems of poor quality and cost — the concept phase of any improvement effort.

However, like many philosophies and tools that are stretched beyond their capabilities, both techniques are encountering difficulty when it comes to their application to front end innovation. While no new idea ever begins with a blank sheet of paper, the front end of any new product or service is hazy and undefined. Six Sigma and Lean tools work well on improving the known, but there must be a product, a process, or some *Thing* to analyze and improve. They are analytical tools that rely on deductive reasoning. They are not inductive tools.

Visualize Hidden Waste

Ford once said, "We swim in a sea of unseen waste. The worst waste is not the machining chips on the factor floor, but the unseen waste of time." Bruce A. Henderson and Jorge L. Larco explain this concept well in their book *Lean Transformation*. Traditional cost accounting processes are generally not lean, either in theory or practice. Most are standard costing systems that may look good on paper, but rarely give an accurate measure of product cost. They draw careful distinctions between direct and so-called indirect labor costs.

Direct labor is usually defined as the people actually working on the product — as compared to maintenance people or material handlers who are considered to be "indirect." Since indirect costs cannot be precisely assigned to labor or material, they are collected in vast overhead pools and a method is devised to spread these indirect costs over the supposedly "productive direct labor." The result is the "peanut butter effect." The causal drivers of indirect cost are lost as they are spread over all products. We end up seeing the effect, but not the cause. It's like a doctor treating your stomachache with some aspirin, without really understanding — and curing — the root cause.

And while accounting practices do a poor job in treating our internal corporate stomach pains, they do little, or nothing, for spotting and curing aches in the external domains of supply chain and customer space.

Attack the Seven Creators of all *Waste*

> If it is bolted, try riveting it. If it is riveted, design it for welding. If it is welded, try bending. If it is bent, try making it flat. If it is thick, make it thin!
>
> —Sign in Ford's Model T Experimental Room

The design of the Model T, and all the transactional processes that surrounded it, testify to Henry Ford's obsession with *Waste* prevention through the use of better front end innovation. Ford's challenge to his team was to find all the opportunities for waste reduction, before they leaped to solutions.

Imagine Norval Hawkins holding his innovation cube in his hand. writing down what he called the real causes of the "stomach aches" of the automotive business. He was providing his team with a "lens" for looking at the current Ford Model N in a completely new way. He used *"waste abstraction"* and gave the Ford team a high enough perspective to see the creators of *Waste* in all their forms.

From...	...To
Complex	Simple
Precise	Adaptable
Variable	Unchanging
Sensitive	Robust
Immature	Proven
Dangerous	Safe
High Skill	Low Skill

Ford's challenge to his Model T team still holds true today. Experience now shows that over eighty percent of all poor quality, high cost and excess time is set in motion by seven types of innovation decisions.

Want the same challenge? Take a piece of paper and list Ford's "Creators of Waste" on the left side. On the right, list an example of each of the seven in your own offering — or that of your competitor. Finding unseen *Waste* can be the start to a breakaway product.

Complexity

This is the use of many process steps and parts when fewer would suffice.

Ford knew every manufacturing process step required time and money, and if done wrong, could lead to a quality flaw. He also knew this was true for the customer in maintaining his auto. The key was to reduce the number and complexity of all these *Ings* or steps. One way he did it was to simplify the Model T product architecture through standardization and modularization. This approach can pay hefty dividends both in the factory, in shipping costs, and in customer repair.

Example: The Model T had forty percent fewer parts than its predecessor, the Model N. Since rail freight was based on space used, not weight, components were shipped in modules for completion at assembly plants.

Ask yourself: What Ings can we eliminate in our value chain to reduce complexity?

Precision

A precision process is one that demands that everything go just right or the entire process will go wrong. Precision is costly in terms of machines, tooling, skilled employees and scrap. Ford was not a great fan of unnecessary precision. He sought the opposite of a precision process. He wanted a "robust" solution, one that is tolerant of wide variation.

Example: Ford believed that when you cannot minimize precision, you must transfer it from human hands to machines. "Let the machinery deliver the precision; let man do the work not requiring precision," he said. Ford was the leader in the use of advanced machine tools.

Ask yourself: What precision Ings can we take out of our value chain? Can we reduce the precision our customer must apply to use our offering?

Variability

In the early auto manufacturing days, a special category of workers were the highest paid and the most in demand — the "fitters." They were the factory floor craftsmen who filed, shaped and hammered parts together. This "craftsmanship" then rolled on into the customer's domain, where an equal amount of skill was needed to replace a part.

Example: Ford set the standard for the Model T with the promise, "We will build the Model T with absolutely interchangeable parts. All parts will be as nearly alike as chemical analysis, the finest machinery, and the finest workmanship can make them. No fitting of any kind will be required." Customers were confident they could get replacement parts that fit the first time.

Ask yourself: How variable are our products, processes or services? Where can we eliminate variability?

Sensitivity

Ford knew the Model T would have to be robust in order to survive the country roads of his day. Most automobiles at that time were low slung, heavy, and never dared venture off city streets without some fear of damage. To emphasize the company's focus on durability, Ford admen entered the Model T into one of the first cross-continent races. The new Model T took first and second place.

Example: Ford raised the ground clearance of the Model T. He reduced its weight and used a new three-point suspension to provide more flexibility for rocking out of snow and mud.

Ask yourself: How can we make our offering more robust? Where are our points of sensitivity to failure?

Immaturity

Ford liked to say there was nothing absolutely new about the Model T, "just a lot of improvements combined in the best way." Other manufacturers would offer new features every year, but Ford saw a high degree of "newness" in a vehicle as being risky. He was willing to constantly experiment with new techniques for manufacturing and selling his Model T. However, he was almost paranoid about changing the basic design.

Example: Ford designers had a new version of the Model T, unknown to him, ready for Ford when he returned from a European trip. When he saw it, Ford went into a fit of rage, grabbed a sledgehammer and completely destroyed the prototype. This resistance to changing anything about his beloved Model T would eventually almost destroy his company.

Ask yourself: What new idea can we use with minimum risk?

Danger

Automobiles in 1907 were considered to be dangerous, both for those riding in them as well as for pedestrians. Steering and braking were major problems. Ford worked on both in the design of the Model T. Up to that time, you could have either right hand or left hand drive for your vehicle. Right hand drive was a favored approach as it let you see how close you were to the edge of the road — or ditch. The Model T changed that approach forever in America.

Example: Ford believed left hand drive was the safest approach to steering. He foresaw the day when two-way traffic would be the real safety problem. He standardized left hand drive and drove costs down due to this common design. Ford now had one single chassis to manufacturer. Other manufacturers soon followed his lead.

Ask yourself: Think of danger as being more than danger to humans. Think of danger to the environment too. Where can we gain a competitive advantage?

Skill Intensive

The key to low cost mass production wasn't the moving assembly line, as many people then and now believe. Rather, it was the complete and consistent interchangeability of parts and the simplicity of attaching them to each other. The automobile boom soon outstripped the pool of skilled labor. Ford knew he would have to design the Model T, and all the transactional processes associated with it, so that unskilled labor could perform them.

Example: After reducing the number of parts in the Model T, Ford then went to work on reducing the skill required to manufacture and service each. His accountant, James Couzens, applied the same tactic by simplifying the purchasing of components. He used single source purchasing to reduce the number of transactions, and promised his suppliers higher volumes if they would lower their prices. Thus, Ford Motor was able to relentlessly drive down the selling price of the Model T every year.

Ask yourself: Where can we eliminate skill in our offering? Are there opportunities in reducing the skill needed to use our service or product?

To identify and attack creators of *Waste*, Ford used the same approach Norvel Hawkins suggested for finding the best Model T values. In this case, Hawkins called it the *Waste Seeker Tool*. Its purpose was to surface and focus attention on the *Ings* the Model T would have to avoid to be a success. He listed all main creators of waste and then how these created problems for the company.

1907 Ford Model N Waste Seeker Tool

Waste Creators	Why?
Complexity	Five different models. Many different parts. Multiple colors. Learning to drive is complicated for new auto user.
Precision	Precision tolerances, especially with engine parts. Hand fitting of parts is necessary.
Variability	Different suppliers with quality, delivery variables between them. Much hand fitting and sorting. Need for standards, common parts
Sensitivity	Low chassis clearance. Difficult for country road driving.
Immaturity	High learning curve with multiple models. Too much experimentation going on all at once.
Danger	Factory injuries too high. Not enough driver protection.
High Skill	Transmission shifting difficult to learn.

Summary

- All value comes at a price. They are the tasks, or *Ings*, that must be performed to deliver that value. Your goal must be to meet your stakeholders' *Wants,* the values they seek, using the absolute minimum of tasks along the entire life cycle of that value.
- Every *Ing* (task) required to create your innovative solution(*Thing)* takes time, hence costs money, and, if performed improperly, will result in a quality flaw. You must apply "fire prevention" rather than "fire fighting" by eliminating as many *Ings* as you can, right at the start of your innovation effort.
- Conventional accounting is a poor guide to waste prevention. It provides guidance on the "visible" costs of labor and material. However, transactional costs and most overheads are lumped together and are almost impossible to separate to find their root causes. The causes of these *Ings* are hidden from your view.
- Seven kinds of solutions create more than eighty percent of all high cost, poor quality, and slow time to market. You can use these seven common waste creators to analyze current offerings for improvement, probe your competitors' products for weak points, and predict "leanness" as you develop your own offering.

Summary of Part One

How to Find Your Best Opportunities

Remember that a successful innovation happens when three paths of knowledge intersect. Think of these as three lines intersecting in a three dimensional space. One is the knowledge of a *Want*. *Wants* may be spoken or unspoken. The best are the unspoken ones that you can see early and exploit. A second path of knowledge is the solution, or what is called a *Thing*. A *Thing* may be something physical, like a product, or it can be a process, a system or a business strategy. A third path of knowledge is your ability to apply your *Insight* to re-shape and make a connection between *Wants* and *Things*.

Chapters 1 through 4 gave you a systematic way to begin finding the right *Wants*. An innovation will be successful when you are able to use your best thinking to find a real *Want* and connect it to a real solution, or *Thing*.

Coming in Part Two – How to Find Your Optimum Solutions

Part Two, Chapters 5 though 7, will give you rules for searching for and finding your optimum solutions. These are the *Things* you need to have to meet the *Wants* you have identified. You will learn a systematic way for developing your *Insight* to do this.

PART II

How to Find Your
Optimum Solutions

CHAPTER FIVE

Search Outside Your Shell

Fatal Flaw: We fail to explore beyond the grooves of our known experience.

How to create breakthroughs... How you can extract new ideas from any experience... Nature has already solved many of our problems... Leonardo's Rule... Insights happen when you explore new territories... How to go off your beaten path to find creative insights

In this chapter you will learn how to move from identifying *Wants*, or opportunities, to how to apply your *Insight* and find real *Things*, or solutions. Don't just see innovation as being "technology." Remember that *Things* are *any* kind of solution. The computer, cell phone and Internet are all highly visible technologies. But innovations in insurance, banking, services, and many other non-technical areas are also solutions.

Grooves of Experience

We spend most of our waking lives thinking in the ruts of our experience. When life presents us with a problem or an opportunity, we "solve" it in accordance with the code of rules that we used with similar problems in the past.

Arthur Koestler, in his classic *The Act of Creation,* compares these ruts to a train engineer who must drive his train along fixed rails according to a fixed timetable. The same ideas, verbal concepts, visual forms, and mathematical entities are used. The train journey is comfortable, seemingly risk free, and without stress. However, there is nothing remotely new or spontaneous about it.

Many times your toughest task is not just coming up with new ideas. It's the difficulty of escaping from the clutches of old ones. We must temporarily discard old solutions and think in a completely different manner. Creative thinking is contrary to our usual pattern of thought. Normally we see a problem, usually well defined, and we systematically apply time-tested theories and tools to solve it. It was Albert Einstein who said, "The problems we face today cannot be solved on the same level of thinking we were at when we created them."

Condition Yourself to Recognize *Insights*

Researchers tell us that we have at least 60,000 thoughts per day. Most of these are along the well-worn grooves of our daily life. An *Insight* happens when you experience a sudden shift in your pattern of thinking, taking you out of these grooves. A new pattern of thought erupts that enables you to "see" something from a totally different perspective. Yet breakthrough *Insights* need not be accidental or mysterious. You can create a mental framework for attracting them.

To understand how to do this, you must first know how *Insights* occur. An *Insight* happens when three other "sights" overlap. The first is *Foresight*, the ability to imagine what might be. Another is *Hindsight*, the experience of the past. And the third is *Outsight*, your ability to stretch your mind beyond the bounds of your present experience to "borrow" new ideas from different places.

Exploiting an *Insight* means the risk of leaving the present for the unknown future. Some are unwilling to make that leap, but Henry Ford did just that. Return with me for a moment to Henry Ford's Experimental Room in 1907.

Henry Ford's Great Leap

"We are going to beg or borrow every good idea from anywhere," Henry Ford abruptly announced. On the Experimental Room blackboard, Norval Hawkins had chalked what looked like a three-bladed propeller.

The room went silent as Hawkins began, "*Foresight* is our ability to imagine what could be, that which does not exist today. I think everybody in this room agrees the time is ripe for a new category of automobile, the car for the common man." Hawkins, forever the accountant, held up the sales numbers for the current Model N. At one point Ford Motor had been unable to meet the demand for this precursor of the Model T.

Hawkins continued, "*Hindsight* is our experience and knowledge, what we know. We are the best in the business." He looked

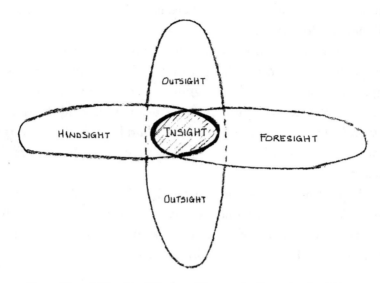

From Norval Hawkins' "Rules of Innovation" notebook, 1907

around the room to see the Dodge Brothers nodding their heads in unison. Alfred E. Sloan also gave a slight nod.

Hawkins went on, "But we need far more. We need *Outsight*. We know our current Model N will not take us to where our *Foresight* tells us we must go. We need new ideas, but new ideas seldom come from just doing our everyday work. We have to apply *Outsight*. This means going beyond your experience and knowledge to explore new ideas from different industries that have already solved the problems we are just now facing. *Outsight* is our 'peripheral vision,' the ability to see things outside our normal experiences."

At this point, Ford jumped up from his grandmother's rocking chair. "It's like riding on the passenger side," he exclaimed. "Driving takes a lot of *Foresight* and *Hindsight,* but you don't have much time to see what's on either side of you. When you are a passenger, you suddenly see new things you have never seen before.

"The air is full of ideas," Ford continued. "They are knocking you in the head all the time. You only have to know what you want, then forget it and go about your business. Suddenly, the idea will come through. It was there all the time."

Insight in Action

Many times the most powerful *Insights* are right in front of us. Take for example the history, not of the car, but of its predecessor the bicycle. The first ones had two identical wheels. The rider sat on a seat and powered it by pushing it along with long strides.

Then somebody had an idea. Why not put pedals on the front wheel? This would mean the rider could go faster with less effort. Then the idea came to enlarge the front wheel for even faster speed and less effort. Since there was no drive train on the bicycle, the only way to make the bike go faster was to make the front wheel bigger.

Bicycle manufacturers began to try to outdo each other. The back wheel was shrunk to only a tenth of the size of the front wheel, which kept growing ever larger. The result was the "penny farthing bicycle," so named for the English coins, one large and one small. Things worked well, although the penny farthing bike was dangerous if you fell, and ladies shied away from riding it.

A tattered photo of one of these early bicycle plants shows all the factory employees standing in front of their lathes. On the left hand side of the shop are the employees manufacturing the large wheels. On the right are all the employees manufacturing the small wheels. Above them whirred the overhead belts that transferred power to drive the lathes, a system that had been in use for decades. It converted the slow rotational power from the water wheel in the sluice through a series of sheaves to the high speed needed for the lathes.

One day an unnamed employee looked up and asked, "Why not use the same drive train idea to power the rear bicycle wheel?" Within only weeks, this safer model with a chain drive and equal-sized wheels replaced the penny farthing and the "safety bicycle" was born. The employee had used *Outsight*, the ability to extract an idea from one experience and use it in another.

Outsight is the Toughest Skill

Every innovation requires some degree of *Outsight*. The more you get out of your shell to explore new disciplines, the faster you will discover ideas to create new *Insights*. You connect bits of your *Hindsight* with your new *Outsights*. You reconfigure ideas to have them emerge as something new. Solutions are assembled from the pieces of the past and the present. Guttenberg took the idea of separate letters from a coin stamping plant to come up with the *Insight* of moveable type. He looked at a wine press the same day and came up with the idea of a printing press.

A World War I military designer saw the art of Picasso and came up with a better camouflage pattern for guns and tanks — a classic case of successfully combining two unrelated things. In an example of similar thinking, the "unbreakable" U.S. military code used in World War II was based on the Navajo language.

It's impossible to think innovatively by looking harder and longer in the same direction. New solutions seldom arrive from the same point on the compass. It was Henry Ford who wrote, "Men have a great love for regularity. They fall into the half-alive habit. Seldom does the cobbler take up with the new-fangled way of soling shoes, and seldom does the artisan willingly take up with new methods in his trade. Habit leads to a certain inertia, and any disturbance of it affects the mind like trouble. Many live in well-worn grooves in which they have become accustomed to move."

Outsight is the most important of the Three Sights. It is not easy to do. Some of us have it naturally, while others struggle to do it. Yet everyone is capable of developing *Outsight* when you treat it as a systematic process.

Crack *Things* Open

We typically never borrow things from other domains in their entirety. We borrow pieces. We then assemble these pieces with other parts to form new *Things* to meet the real *Wants* people seek. The new is built from the pieces of the present and the past. When you are able to break apart the familiar things of life to examine them in a new way, you can find new ideas.

Take for example, the story of White Out™, sometimes called "liquid paper." You would think it originated in a paper company lab. Not at all. White Out began with a very clear *Want*. Bette Nesmith, a typist, was the inventor of liquid paper. Bette was working as a secretary when she wondered why artists could paint over their mistakes but typists could not. Using her kitchen blender, Bette mixed up a batch of water-based paint to match the company stationery and poured it into an empty nail polish bottle. She then took it to work. Using the small brush, she could paint over and fix typos. White Out was born. She had borrowed an idea from another domain to solve her problem, as well as created an entirely new product. Gillette later bought her company for $48 million.

Creative ideas can just as easily come from nature. The winged structure of an elm tree's seeds inspired the engineering applications of an improved windmill design, safer helicopters, and improved ski design. The fly's vertical take-off inspired vertical take-off aircraft, while the burdock burr's hooked spines provided the basis for Velcro fasteners.

Every product, process or system we work with in our everyday lives has ideas we can adapt to our own challenges. The hurdle, though, is that we have no way to systematically break *Things* apart to find ideas we can use. *Things* outside our normal scope of experience do not give up their secrets easily. You must have a method for quickly breaking them apart to look for the ideas within. Without a systematic way for doing this, you can wander aimlessly, looking for ideas that don't really meet your purpose. You can stumble over a diamond in the rough without recognizing what it is, pick yourself up, brush yourself off and continue on your way as though nothing happened.

How to Break *Things* Apart to Find Ideas

The secret to finding new ideas is knowing how to do it. Think of *Value*, what our customers really want, as being made up of five core elements. You can break anything into these five parts and systematically probe each for ideas you can use.

Let's return again to Ford's Experimental Room, where Norval Hawkins has chalked a large box divided into five parts, much like a puzzle.

"To achieve our Model T goals we have to think differently," Hawkins says. "Just copying what others are doing won't give us an advantage. Everything is the sum of five elements, and it doesn't matter whether it is your home, your bicycle or your insurance policy. When we break apart *Things* into these five elements we can

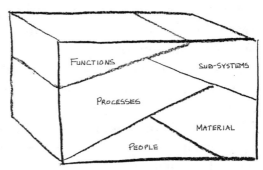

Norval Hawkins' Value Puzzle.
From Norval Hawkins' "Rules of Innovation" notebook, 1907.

begin to see how other industries are using each of them."

Hawkins continues, "We have to use *Outsight* to arrange these pieces in different ways. We must look at what other industries are doing and learn from them."

Here are the questions Hawkins challenged the Model T team to think about as he reviewed the Five Elements of his *Value Puzzle*.

1. Functions. These are the "purposes" of any product or service. They collectively define *Value*. Are there any new functions we can adopt from another industry?

Functions are the "jobs" that your offering delivers. Think of your car. It conveys you and your family from point to point. It also protects you from the environment (rain and snow) as well as other cars that might crash into you. Your car also provides you an "image."

Ford stripped all but the barest functionality from the Model T. While other manufacturers were beginning to add features and styling, the Model T went in the opposite direction. The Model T's functions basically stayed the same for the nineteen years it was manufactured.

Functions are best expressed as a verb and a noun. There is no product, service or business transaction that has only one function.

How to Start Looking: Select a product or service similar to yours but not in your competitive space. For example, if you are in the banking field, look at the insurance business, or vice versa. Use a verb and noun to list all the functions that adjacent field is providing with its service. Then ask yourself, "Can we borrow some functional ideas?"

2. Sub-Systems (Parts). These are the elements that embody and deliver *Functions*. Are there any sub-system ideas we can adopt from another industry?

Think of your car as the sum of "hard" sub-systems such as the drive train, engine, body, and many more. However, it is also made up of "soft" sub-systems such as your warranty, payment or lease plan, and service parts policy.

Some of the greatest innovations in the auto industry have been on the "soft side." Ford deliberately designed the Model T as a group of sub-systems for easy outsourcing. The Model T team then took a sub-system view to develop systems for purchasing, marketing, and manufacturing. The Model T was a "system of systems," both "hard" and "soft." This enabled it to be built on a massive scale not known before in any consumer industry.

How to Start Looking: Break your product or service into the sub-systems that comprise it. For example, Dell Computers did this in its early days and copied outsourcing and assembly sub-system ideas from a wide array of industries.

3. Materials. Think of materials as being everything that is needed to create and maintain the *Sub-systems* described above. Are there any ideas we can borrow from adjacent businesses?

Ford saw materials as being not only the steel and wood used to build the Model T, but the fuel and oil to operate it. In a period when the *New York Times* was reporting that it cost $1,500 per year to operate an automobile, Model T owners could boast of annual operating expenses of less than $100. A well-kept Ford was said to cost about a penny per mile to run — or as one thrifty fellow noted in the March 1912 *Ford Times* magazine, one-fourth of a penny when he took his entire family along. Ford borrowed the idea of a new, lightweight steel called vanadium from a French car manufacturer. This enabled him to reduce weight without compromising strength. The result was a more durable auto with higher speed and lower horsepower.

How to Start Looking: Materials are fertile ground for idea exploration, both in their hard and soft versions. Most project teams do a good job of recognizing and improving *Sub-Systems* and *Materials*. Most are visible. We have corporate measurements for them and innovation teams are held accountable for improving them. Are any new materials being used in industries adjacent to yours?

4. Processes. These are the "jobs to be done" to convert *Materials* **into** *Sub-Systems* **that deliver** *Value.* These "jobs to be done" are both "visible" and "invisible." Visible processes are the *–Ings,* such as manufacturing, assembling, packaging and many others. The invisible processes include inspecting, checking, delivering, repairing, among others. Henry Ford was an intuitive *Ing* thinker. He had the ability to "see" these "jobs to be done" where others did not.

How to Start Looking: The lean manufacturing revolution in the American auto industry began when it recognized that Toyota was attacking the "hidden" process costs on the auto factory floor. Dell Computer's success in the computer market was driven by a focus on the process of delivery, driven by a streamlined supply chain. Ford borrowed the concept of the moving assembly line from the moving *disassembly line* of a Chicago meat packing plant.

5. People. People are fundamentally at the core of any offering, and their skills are a part of your *Value Chain.* **What businesses are applying peoples' skills in a different way?**

Ford believed the Model T was all about people. "You start with the value that people want, and you work all the way back," he once said. This is the last and most important of the *Five Elements* to explore outside your domain of normal experience. People must perform all the processes required to deliver value.

How to Start Looking: Ford integrated machines with men. He used the most advanced machine tools to provide precision, and relied on people to maintain, load and unload those machines. The current trend of outsourcing "back office" processes to India is an example of how companies are utilizing cheap communication to use people in a new way.

Norval Hawkins' Outsight Tool

The following are the three industries the Model T team first explored for new ideas. Norval Hawkins captured these *Outsight* ideas in his notebook.

1907 Trends	Bicycle Industry	Food Industry	Mail Order Home Construction Industry
Functions (Purposes)	Freedom to go anywhere, anytime	On-the-shelf, immediate use of product	Immediate delivery, no waiting
Sub-Systems (Parts)	Standardization of wheels, bearings, brakes	Emerging government standards	Modular construction, order from Sears & Roebuck
Materials	Lighter weight, stronger	Cheaper, lighter, thinner	"Balloon" design to reduce shipping weight and cost
Processes	Precision machining, modular build	Food production as a continuous process	Pre-cut components, segmentation of labor
People	Low cost, mass market, transportation freedom	Integrated value chain from farm to consumer	Affordable housing for the mass market

Summary

- "Eureka!" moments, or *Insights*, occur when your *Hindsight*, *Foresight* and *Outsight* converge. *Hindsight* is your experience of the past. *Foresight* is your imagination of the future. And *Outsight* is your ability to extract new ideas from different fields of knowledge.

- *Outsight* is by far the most difficult of the three skills. You can improve your *Outsight* by breaking any product or process into its five basic elements. This trick of "cracking open" anything reveals new ideas. When you change your way of looking at *Things*, the *Things* you are looking at change.

- Nature has already solved many of our problems. Start your search for new ideas there. How has Nature done the task? What plants, animals, insects and more can you look at for inspiration?

- *Insights* seldom happen when you are in familiar circumstances. They arrive when you explore new territories. All humans have the capacity for creative activity. However, this is often suppressed by automatic patterns of thought and routine behavior of daily life. Breaking out of the ruts of daily life by using the *Outsight* tool described in this chapter will enable you to see *Things* as you never have imagined them before.

CHAPTER SIX

Connect *Things* in a New Way

**Fatal Flaw: You have no systematic way to
combine new ideas into new solutions.**

*How to go beyond traditional "brainstorming"
to focus your idea search for effectiveness and efficiency... The important Rule of Can We?... Since
we tend to use the same problem-solving approaches repeatedly, we usually come up with the same
answers... How to find new solutions by looking
at your opportunity in many different ways, yet
without losing strategic focus...New solutions
emerge re-combining both old and new ideas...
How to systematically combine different ideas to
create new solutions*

Creative thinking is like a salad bowl. It has you mixing old
ideas with new ones to come up with a wonderfully
improved recipe. You may not remember how you did it,

but you know it happened. All decisive advances in the history of scientific thought happened the same way. Breakthroughs come from mental cross-fertilization between different disciplines. The mind connects different things in a new way. Think of your world in this way: the future is already here. You have just not found all the ingredients and added them to your salad bowl.

Find Many Ideas

A great innovation begins with a lot of ideas. It involves combining and recombining ideas, images, and thoughts into different configurations. This happens in our conscious as well as subconscious mind.

Consider Einstein's equation, E=mc2. Einstein did not invent the concepts of energy, mass, or speed of light. Rather, by combining these concepts in a novel way, he was able to look at the same world as everyone else and see something different. Einstein vaguely referred to the way he thought as "combinatory play."

Ideas start with the recognition that a gap exists between a *Want*, whether hidden or open, and a *Thing* available to meet that *Want*. A lack of good ideas is rarely the problem in a company striving for growth. The challenge lies in selecting the good from the bad and then shaping them into a real product or service.

Ask "Can We?"

Since we tend to use the same problem-solving approaches repeatedly, we usually come up with the same answers. The Rule of "*Can We?*" forces us to look at our problems in a way we might not have done.

Ideas will begin to sprout as you read through the following. Jot them down in the space provided or use a clean sheet of paper. At the end I will show you how to use these ideas to create even more.

Let's go back now to the Experimental Room as Hawkins returns to his *Value Puzzle*. He jots down eight questions and challenges for the Model T team to apply. Rubbing out part of the *Functions* piece of the puzzle, he begins, "A non-existent part, process, or task takes no time, hence costs no money and, should it be flawed, will not result in a quality problem! Think about what *Things* we can eliminate."

Can We <u>Eliminate</u> Some *Thing*?

What Ford Said: Henry Ford told the Model T team, "We will start with an assembly that suits the function of the Model T. Then we will find some way to eliminate the entirely useless parts. This is simple logic, but oddly enough the ordinary process starts with a cheapening of the manufacturing instead of with a simplifying of the article. First we ought to find whether it is made as it should be — does it give the best possible service? Then, are the materials the best or merely the most expensive? Then, can its complexity and weight be cut down? And so on."

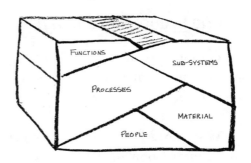

From Norval Hawkins' "Rules of Innovation" notebook, 1907.

Jump Start Your Thinking: When you are getting your morning coffee and donut, think about this innovative idea. For over two hundred and fifty years, donuts lacked a hole. Their original size was about that of a walnut, with no hole in sight. The Pilgrims learned about the pastry during their stay in Holland and brought them to New England. They called them "dough nuts."

Then in 1847 Hanson Gregory, a sea captain from Maine, had an idea. One evening Gregory came on deck to see his first mate manning the helm with a donut hanging on one of the spokes of the ship's wheel. The mate wanted to have his donut close at hand.

When Gregory went below to get his own, the cook told him the new donut with the hole in the center cooked more evenly, hence tasted better. Captain Gregory took the idea home to his mother who owned a shoreside restaurant, and a new American pastry was born. Today his name and contribution are commemorated in his hometown of Rockport, Maine, with a bronze plaque.

Can We <u>Reduce</u> Some *Thing*?

Erasing part of the *Parts* puzzle piece, Hawkins continues with his next challenge, "If we can't eliminate it, can we *reduce* some *Thing*?"

What Ford Did: Almost every year from 1909 to 1925, the Ford Motor Company was able to reduce the price of the Ford Touring Car. The price slid from $950 in 1909 down to an astounding $240 in 1925 — for an even better car. While prices of other cars were going up, the Ford came down and down. No other car-maker could match Ford.

The company could have added an electric start to the Model T after the device was intro-

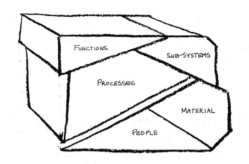

From Norval Hawkins' "Rules of Innovation" notebook, 1907.

duced in 1912. But Henry Ford did not want to install the battery needed to power it. Adding a battery meant more cost and, very importantly, more weight. Ford Motor finally offered a factory-installed, battery-powered electric starter in 1921. Until then, owners had to get starters installed as an aftermarket item.

Jump Start Your Thinking: Automotive designers struggled for years to improve the performance and quality of tires, while constantly battling steadily increasing material costs. An automobile's tires are a major supply chain expense item. The designers used conventional thinking to strengthen tire walls and treads, as well as finding new material combinations. Then one unconventional tire team applied innovative thinking. Why not go in the opposite direction? They focused on the "fifth" or spare tire, seldom used today. They used the technique of *repurposing*.

They first looked at the spare tire's *function*. Instead of being viewed as a replacement, the tire's purpose was shrunk to be only a *temporary* way to get you to where you could fix your tire or buy a new one. The result is the small spare you see in your trunk today. Performance was intentionally reduced. The size was minimized, thus reducing material. It became lighter and easier to handle during a tire changing crisis. And it requires less space in your trunk. Very importantly, the cost was reduced more than fifty percent.

Can You Substitute Some *Thing*?

Turning to the blackboard again, Hawkins challenges the team to look at *substitution* as an innovation technique.

What Ford Said: Henry Ford disliked the use of wood because of its innate inefficiency. "Some day we shall discover how further to reduce weight.For certain purposes, auto bodies for example, wood is now the best substance we know, but wood is extremely wasteful. The wood in our automobiles now contains thirty pounds of water. There must be some way of doing better than that. There must be some method by which we can gain the same strength and elasticity without having to lug useless weight." Ford introduced the use of lightweight vanadium chrome steel to eventually replace wood as a body frame.

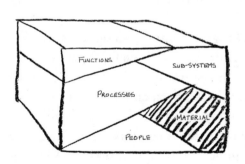

From Norval Hawkins' "Rules of Innovation" notebook, 1907.

Jump Start Your Thinking: The American auto industry was one of the first businesses to see the benefit of engineered plastics. Plastics enabled parts to be lighter weight. They also enabled several functions to be integrated into one part. The substitution of injection molded plastic for steel reduced the process steps needed for production. Substitution is also going on in the transactional process of buying and selling cars. The Internet is being substituted for face-to-face purchasing.

Can You <u>Separate</u> Some *Thing*?

Hawkins moved to his next challenge, "Can we *separate* one or more of the pieces of the puzzle to gain an advantage?" Redrawing the materials piece, he challenged, "Can we separate the Model T into sub-systems for assembly nearer to the customer?"

Separation is the tactic of dividing any of the five elements into smaller "chunks" that can be easier to outsource, assemble or maintain in the field. With transactional processes, it is the idea of breaking large, complex and cumbersome steps into smaller, simpler ones that can be better automated and delivered.

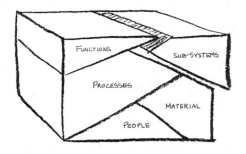

From Norval Hawkins' "Rules of Innovation" notebook, 1907.

What Ford Did: The Model T had a new engine, along a design previously thought unworkable. The top part, or head, was removable. This meant easier servicing, without having to remove the entire engine. Norval Hawkins came up with another idea. He asked, "Can we ship separate components to our distributors for their assembly to cut down on railroad shipping costs?" The result grew into the technique of building major components in Detroit and then shipping them to assembly plants across the country. Freight costs were cut byseventy-five percent.

Jump Start Your Thinking: The birth of the flashlight came from the idea of separation. When Russian immigrant Conrad Hubert arrived in New York in the 1890s, he took a job with Joshua Lionel Cowen, the man who would one day create Lionel trains. Cowen had just perfected what he called the "electric flowerpot." It consisted of a slender battery in a tube with a light bulb at one end. The tube rose up through the center of a flower pot to light up an artificial plant. Hubert believed in the commercial potential of the electric flowerpot and convinced his boss to sell him the patent rights. But when the novelty item failed to catch on, Hubert was left with a large overstock.

In an attempt to salvage his investment, he separated the lights from the pots, lengthened the design of the cylinder, and received his own patent for a "portable electric light." The light sold so well that Hubert named his new company Eveready Flashlight Company. When he died in 1928, Hubert was able to leave six million dollars to charity. He had applied *separation thinking* to his failed flowerpots.

Can We <u>Integrate</u> Some *Things*?

Hawkins challenged the Model T team to examine all five parts of the *Value Puzzle* and ask the question, "Can we integrate any of the five to create more *Value*? Can we integrate the sales process to make it easier for customers to buy a Model T? Can we integrate purchasing processes to eliminate paperwork? Can we integrate parts of the auto to make them stronger, yet easier to manufacture?"

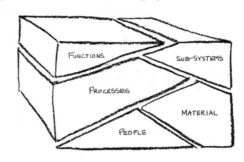

From Norval Hawkins' "Rules of Innovation" notebook, 1907.

What Ford Did: The Model T engine was ahead of its time. The four cylinders it enclosed reflected a major innovation. All were cast from one block of metal. These changes made the cylinder block stronger and lighter, yet easier to service. Another integration idea was encasing the rear axle in a vanadium steel housing with compensating gears in the center. This made for better reliability on bad roads. Yet another integration tactic enclosed the motor, transmission, and flywheel so that their lubrication system could be shared. The Model T was one of the strongest — if not *the* strongest — cars in production in 1908.

Jump Start Your Thinking: Ice cream has been around since 2000 B.C. and is now rated America's favorite dessert. But it wasn't until 1904 that the ice cream cone was born, at the St. Louis World's Fair. Working side by side in one food area was a Syrian baker, Ernest Hamwi, specializing in waffles, and a French-American ice cream vendor, Arnold Fornachou. The story goes that the ice cream vendor ran out of paper ice cream dishes and rolled one of the waffle maker's waffles into a cone. The result was an immediate hit and a World's Fair sensation. Only ten years after the World's Fair, over one-third of all the ice cream consumed in the United States was eaten atop cones.

Can You <u>Re-Use</u> Some *Things*?

Next, Hawkins challenged the group with, "Can we re-use existing, proven solutions from our present Model N in the new Model T?"

What Ford Did: Henry Ford was proud to say that most of the systems in the new Model T were just improvements of the existing Model N solutions. For example, the dual braking system of the Model N was identical to that incorporated in the Model T.

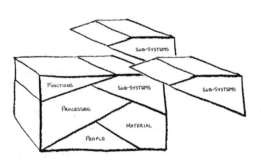

From Norval Hawkins' "Rules of Innovation" notebook, 1907.

Jump Start Your Thinking: On November 30, 1903, a little more than two weeks before their epic flight, the Wright Brothers faced a dilemma in Kitty Hawk. They were having trouble securing the sprockets to their propeller shaft. Orville tells how they solved the problem: "While in the bicycle business we had become well acquainted with the use of hard tire cement for fastening tires on the rims. We had once used it successfully for repairing a stop watch after several watchsmiths had told us it could not be repaired. If the tire cement was good for fastening hands on a stop watch, why should it not be good for fastening the sprockets on the propeller shaft of a flying machine?"

It worked. Seventeen days later, on December 17, 1903, the cement held up well enough to enable the Wrights to make four flights against twenty-one mile per hour winds, with an average air speed of thirty-one miles per hour and the longest flight on record of fifty-seven seconds.

Can We <u>Standardize</u> Some *Things*?

What Ford Did: It was John Dodge who came up with the suggestion of standardizing on left hand drive. The Dodge Brothers built the steering mechanisms; some were left drive, others right. The Model T was the first product model to feature standard steering on the left. When traveling on dirt roads with ditches, drivers preferred to steer on the right

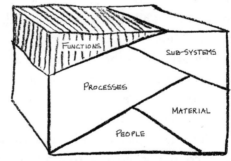

From Norval Hawkins' "Rules of Innovation" notebook, 1907.

so that they could see how close they were to running in the ditch. However, Ford foresaw the advent of two-way traffic and standardized on left hand steering so the driver was better placed to watch out for oncoming traffic. All other manufacturers soon followed suit.

Jump Start Your Thinking: The container ship is a classic case of standardization. In the early 1950s the ocean-going freighter was believed to be growing obsolete. The idea was that it would be replaced by airfreight except for bulk commodities. The costs of ocean freight were rising at a fast clip and it took longer and longer to get merchandise delivered by a freighter as one port after another became badly congested. This also increased pilferage at the docks as merchandise piled up, waiting to be loaded, when vessels could not make it to the pier.

The basic reason for this was that the shipping industry had misdirected its efforts for many years. It had tried to design and build faster ships that required less fuel and smaller crews. It concentrated on the economics of the ship while at sea and in transit from one port to another; however, a ship is capital equipment and for all capital equipment the biggest cost is the cost of not working.

Everybody in the industry knew that the main expense of a ship is interest on the investment. Yet the industry kept on concentrating its efforts on costs that were already quite low: the costs of the ship while at sea.

Another solution, which in retrospect seems simple, arose — to uncouple or separate the stowage from the loading. The idea of a container is to do the loading on land — in fact, at the very point of manufacture, where there is space and time for it to be performed — before the ship reaches port. All that has to be done in port is to put on the container and remove the preloaded freight. This solution was focused on the cost of not working rather than on the cost of working. The answer was the roll-on/roll-off loading of the container ship. The result of this stan-dardization has been startling: freighter traffic in recent years has increased up to five-fold or more, costs overall are down sixty percent or more, and port time has been cut by three quarters in many cases — and with it congestion and pilferage.

Can We <u>Increase</u> Some *Thing?*

What Ford Did: It was Henry Ford who asked the ques-tion, "Can we increase the distance from the ground to the bottom of the chassis to make the T adaptable to the muddiest, rockiest, snow filled roads in America?"

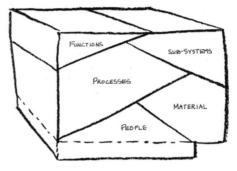

From Norval Hawkins' "Rules of Innovation" notebook, 1907.

In 1908-1909, more than eighty percent of the roads used by automobiles were neither paved nor graded. Many were merely ruts in the dirt. The Model T was ready for them. It was built high off the ground, giving it what one person called a "clunky silhouette," but that kept its belly out of the mud. It also had a three-point suspension that allowed the chassis to twist. By doing so, as the wheels pulled the car out of a rut, they would stay in contact with the ground. As a joke ran about the man trying to explain why he wanted to be buried in his Model T: "I ain't never seen a hole yet she couldn't get outta."

Ford Model T "Seed Bed"

Norville Hawkins described the "Can We" questions as a "seed bed" for ideas.

Can we...	Functions	Sub-Systems	Materials	Processes	People
Eliminate?					
Reduce?			Reduce Model T weight by using vanadium steel?		
Substitute?					
Separate?		Assembly into modules for more compact shipping & assembly nearer to the market?			
Integrate?					
Re-Use?					
Standardize					Offer only left hand steering for people to simplify manufacturing, ordering and learning?
Increase?					

Breakthrough innovations come from breaking apart and recombining existing knowledge in a new way. Arthur Koestler, in his classic book *The Act of Creation,* states this well: "The creative act is not an act of creation in the sense of the Old Testament. It does not create something out of nothing. It uncovers, selects, re-shuffles, combines, synthesizes already existing facts, ideas, faculties and skills."

The first purpose of Norval Hawkins' idea "seed bed" was to create many new ideas. The second purpose was to link ideas in new ways. Notice that his "seed bed" has the possibility of forty ideas. However, when you apply this template to each of the eight *Ilities*, or *Values* they chose for the Model T, they had a matrix of 320 "seed bed" ideas. Add the seven *Ing* Wastes to these eight *Ilities* and you have the possibility of 600 ideas!

The next step is to connect these "seed bed" ideas in new ways. You can do this by using the framework of the five elements, *Functions, Sub-Systems, Materials, Processes,* and *People.* Doing this manually is a daunting task. However, using the digital InnovationCUBE described in Chapter 9 makes it an easy one.

James Burke, in his series *Connections,* says that new innovations are simply the rearrangement of old or existing ideas in a new way. He finds that such recombination should be no surprise, as it conforms to the operating mode of the human brain. The human brain is composed of one hundred billion neurons, each linked by up to fifty thousand dendrites, each capable of linking to fifty thousand other dendrites, indicating a possible total number of ten trillion cerebral connections. This means that the number of ways a signal (a thought) can go in the brain is greater than the number of atoms in the universe. All innovations are the result of connecting *Things* differently.

German auto engineer Wilhelm Maybach did just that. He put together the idea of his wife's perfume spray with the need for spraying gasoline into a carburetor and came up with a reliable system.

Think End-To-End

It was Thomas Edison who said, "All parts of the system must be constructed with reference to all other parts, since, in one sense, all the parts form one machine." Keep in mind that your total project solution depends more on how each sub-solution interacts with other solutions, rather than on how these solutions act independently. When one project solution is improved independent of another, the total project solution can actually suffer. That's why problems and solutions (*Wants* and *Things*) must co-evolve in the innovation process. You must develop them in parallel, sometimes leading to a creative redefinition of the problem or to a solution that lies outside the boundaries of what was previously assumed to be possible.

And remember that bananas do not all ripen at the same time. New technologies do not happen in a straight line; there are many pauses, turns, and blind alleys along the way. When Johann Gutenberg invented the printing press with loose type in the fifteenth century, he cleared the way for the rapid growth of the printed word. It would seem an obvious second step to use the printing technology to create a machine that could take over the cumbersome task of writing by hand. The engineers and inventors of the day were clever enough. A clock, after all, has a mechanism that would seem far more complex than that of a typewriter. So why didn't they build one until 400 years later?

The answer is simple enough. There was no need for typewriters in a world where cheap labor was abundant and where machines were expensive. It was not until the nineteenth century, when industrial production was automated and boomed, that the time was ripe for the typewriter.

New technologies are especially notorious for the time it takes for them to ripen. Fuel cells were first demonstrated in

1839 and have been promoted as an energy solution for decades, but they still have a long way to go to be viable in the mainstream market. Solar cells are another case of slow ripening; the photovoltaic effect that underlies their operation was first noted over 150 years ago. Batteries have also been a slow mover since their invention in 1799. More than 200 years later, a modern AA alkaline battery holds less energy than that in a gram of sugar, and costs fifteen thousand times more per energy unit than electricity from your local power company.

Every innovation must "fit" into the existing world. Thomas Edison knew that. He didn't just invent a product; he also invented a *business system*. The product was the light bulb, but the innovation was the business system. You do not develop a product in isolation from the system, and you do not develop a system in isolation from the product.

Failures in innovations happen when you focus only on the product and ignore the system — for example , the London inventor Swan, who created the light bulb at the same time Edison was developing his idea. While Edison's team at Menlo Park was performing experiments on the incandescent bulb, Edison himself was developing the business system around the light bulb. He was finding answers to system questions. These were the daunting tasks of getting electrical lines into offices, how he was going to provide financing, what kind of fixtures he would use so people wouldn't be frightened to have electricity in their homes, and so on. Swan did nothing about his business system; he only worked on the light bulb itself.

In the U.S., cities were using incandescent lighting a year after Edison perfected the bulb.London didn't begin to use electric lighting in the city until the turn of the twentieth century. It continued to rely on gas lights. The lesson is that we have to be thinking in the spheres of business systems, processes, and product simultaneously.

Toyota followers call this practice "set-based" design. Toyota has one of the most successful product development organizations in the world. Academics and practitioners alike attribute much of their success to what is called "set-based concurrent engineering." This is the practice of looking at many different kinds of solutions, or "sets," before making a final choice. Many solution sets are carried well into the prototype stage before this final selection is made. As an innovation leader, never be satisfied with one initial solution.

The practice of set-based engineering begins by clearly mapping the solution space and broadly considering sets of possible solutions. These are gradually narrowed to converge on a final solution. Casting a wide net from the start and then gradually eliminating the weak or risky solutions makes finding the best option more likely. At first glance you may think this means unnecessary expense, but Toyota doesn't think so. The prototype designs that are not selected may be usable at a future date, or some of the knowledge gained from them can be very useful in terms of what not to do. Knowledge has a high value at Toyota. The data gained from such advanced prototyping is carefully documented and made available to all engineers.

Remember that all solutions, no matter how ideal they may seem at the time, are merely transitional steps. The Three Sharks of change are constantly shifting marketplace, technology and competition to alter your original purpose and possibilities. So all innovations must be formed with the probability of change in mind.

Summary

- The route to great ideas is finding many ideas. Think of how to go beyond traditional "brainstorming" to focus your idea search for maximum effectiveness and efficiency.
- Since we tend to use the same problem-solving approaches repeatedly, we usually come up with the same answers. Consider how to find new solutions by looking at your opportunity in many different ways without losing strategic focus.
- New solutions emerge from mixing old and new ideas. Think of how best to systematically combine different ideas to form new solutions.
- Think end-to-end. See the total system you must build to be successful.

CHAPTER SEVEN

Measure to Learn

Fatal Flaw: You don't apply measurement in a positive way.

Why you must measure to learn, not to "prove"... How to build your measurement system... How you can measure what is meaningful, not just what is easy to measure... Make sure you are headed in the right direction first; think about precision later... Avoid sub-optimization... Measure both your Ilities and Ings... Countering measurement excuses

The purpose of this chapter is to show you how to create a meaningful measurement system at *the very front end* of your project. Too many leaders stumble because they do not clearly understand, or communicate well to their team and management, how measurement works and the value of doing it.

Past Tense Metrics Not Helpful

Feedback is the primary benefit of measurement. Feedback tells you how to make course corrections to assure your innovation strategy is on the right course. This feedback, however, must be in *real time*. You don't want to learn, after losing the battle, where you *should* have gone.

Unfortunately, most corporate innovation metrics don't do this. They are what I call "post process" measurements. They are like the World War II reconnaissance plane taking photos after a bombing raid. In direct contrast, more modern weaponry gives feedback for course correction. The missile heads in the general direction of the target. It then gets more feedback *along the way* to enable it to precisely enter a target's doorway. This is the kind of measurement system you must have to succeed.

Running a project without real time measurement can be dangerous to the health of your organization .By "real time" I mean measuring how well you are doing while your project is underway. I don't mean "past tense" measurements that show how well your product performs once it is out in the marketplace. Neither do I mean measurements that show quality and cost performance on your factory floor.

These metrics are what is known in the business as "lagging" indicators. They describe the past tense. They are not predictive. While they are useful for analysis of past work, they offer little or no guidance for you in your struggle to go to a place no one has been before.

How to Measure

I am now going to show you a way to build your own measurement system starting on Day One. It will give you real time feedback on finding your innovation "sweet spot" and enable you to measure the gap between the *Wants* and the *Things* you are trying to connect. And it will sharpen your *Insight* to help you make that connection faster.

But I must give you fair warning. You will initially encounter a lot of resistance to the method I'm about to describe. Why? The fact is that most people do not understand the basic concept of innovation measurement, both in terms of its limits and its possibilities. You will find that we all carry with us a paradigm for measurement. However, most of these paradigms are not applicable to the front end of an innovation effort. For example, in the scientific and engineering world, measurement is all about precision and the application of instruments to do the measuring. These extremely precise instruments remove variation and human judgment. The *Things* that are being measured are well understood and the metrics for doing the measuring are known to all. These are what I call "hard measurements." I say that with tongue in cheek because, in reality, they are the easiest measurements to perform.

In your battle for innovation supremacy, you must be measuring how well the *Wants* you are aiming for are being met by the *Things* you are developing. The purpose is to give you more *Insight* into how best to reach your *Innovation Sweet Spot.*

Three Data Points

All forms of measurement require three data points. The same is true of the InnovationCUBE measurement system. The first data point, and the one most commonly understood, is the need for a *goal*. A goal is a destination. While innovative project goals may be fuzzy at first, they still are needed in order to give you directional guidance. In point of fact, it is best that they are fuzzy. Rigid goals never work since innovation is a constant back-and-forth journey from defining the right *Wants* to finding the right *Things*, or solutions. Rigid goals can drive you into blind alleys with no way of escape.

The second data point is your place of departure. This answers the question, "Where are we now?" For example, imagine you are flying to Los Angeles from New York. You could know the precise GPS coordinates of Los Angeles, your goal, but if you did not know the coordinates of New York, your departure point, you might head east rather than west.

The third data point is where you are in your transit from your departure point to your destination. This tells you whether you are moving forward, backward or sideways. None of these three points need be "precise" in the early stages of your innovation challenge; they need only be *directional*. It is more important to know that you are generally headed north than whether you are precisely on track.

What You Must Measure

I started out this book by explaining that a successful innovation happens when three paths of knowledge connect, leaving no gap between them. This is when your sweet spot of innovation occurs. But how can we know we are moving in the right direction with all three paths?

You must use measurement as an *Insight* tool for finding new *Wants* and delivering the best *Things*. That is why you must measure what is important — not just what is well known and easy to measure. We often spend most of our time measuring those things we understand well and can measure them precisely, yet these are not the ones that are most important for success.

You must first determine the conditions for success, the *Wants*, and then use your *Insight* to try various *Things* to bring them in alignment with the *Wants*. Along the journey, there will be many twists and turns. What can help keep you on track is the application of measurement.

Author Roger von Oech points out that some *Things* don't come into existence unless the right conditions are brought together. Solar eclipses only occur when the sun, moon and earth are aligned. Rainbows only occur when the sun is behind the observer, less than forty-two degrees above the horizon, and shining into the rain. Political revolutions only happen when there is an economic downturn, vast social inequity, and an attractive ideology. Likewise, you must find the right combination of *Things* that will align with the right *Wants* to create an *Innovation Sweet Spot*.

As you learned in Chapter 3, the best place to start discovering the right *Wants* is with the eight fundamental values, or *Ilities*, that all customers seek. Because all ideas start with the recognition of gaps, all processes for identifying great ideas must be aimed at creating perspectives that make the gaps and holes visible. Apple saw gaps in the way people acquired and used music (accessibility and use-ability) and created the iPod and iTunes. Southwest Airlines saw "affordability" gaps in the price of service for airline travel, and designed an approach that delivered no-frills service at significantly reduced fares.

That which gets measured, gets done. We all perform as we are measured. Using wrong measurements will deliver wrong results. A doctor who incorrectly diagnoses a disease won't prescribe the right treatment.

Measure *Wants* to Find Gaps

Because all ideas start with the recognition of gaps, all processes for identifying great ideas must be aimed at creating perspectives that make the gaps and holes visible.

The reason why so many leaders do not use innovation measurement is that they simply *do not know where to start.* They measure what is easy and well known, not what is meaningful. Measures must be linked to the factors needed for success, the key business drivers. Fewer are better. Concentrate on the vital few rather than the trivial many. Align your project metrics with what you want to deliver to your customer and yourself. Identify your stakeholders then consider measuring what they need to meet their needs.

The *Wastes* of quality, cost and time can be identified and proactively attacked at the very front end of any project. The InnovationCUBE method described in Chapter 9 begins with you looking at an existing product, or one of your competitor's, through the lens of the seven creators of waste described in Chapter 4. Measure *Things* to see your progress and make decisions. Measure to see where you are, then improve it again.

Measure *Things* to Discover the Best Alternatives

Innovation is the commitment of present resources to future expectations. The value of a company's stock placed on it by Wall Street is the commitment of present dollars to future expectations. That means opening your organization to uncertainly and risk. In order to minimize uncertainty and risk, we must have some sort of measurement to ensure we are at least headed in the best direction.

David Kelly of IDEO once said, "Enlightened trial and error outperforms the planning of flawless intellects." What he was advocating was the technique of "failing fast." This is the rapid prototyping of ideas to determine whether they have validity.

Successful leaders "balance" their product development efforts. They identify the key success factors and make sure that one is not optimized to the detriment of another. You must take a *systems approach,* optimizing the total product and not just one success factor.

Who Should Select the Metrics and Do the Measuring?

In the introduction to his bestselling book, *The Wisdom of Crowds,* author James Surowiecki describes an event in the fall of 1906 that began to get people thinking differently about group intelligence. During that time, the prevailing opinion was that only a few people had the characteristics to make good decisions.

It seems the British social scientist, Francis Galton, was visiting a country fair. Galton had spent most of his career trying to isolate those decision-making characteristics to prove that the vast majority of people did not have them. Only if power and control stayed in the hands of the select, well-bred few, Galton believed, could a society remain healthy and strong. As he walked through the fair that day, he came upon an ox weight judging contest. All sorts of people, young and old, farmer and non-farmer, were lining up to guess the weight of the ox after it had been "slaughtered and dressed." Winners would receive prizes.

Galton joined the other eight hundred people who tried their luck. Then he seized upon the opportunity to prove his political theory that the "average person," like the "average voter," was not as capable as the "expert." When the contest was over and the prizes awarded, Galton borrowed the tickets and ran a series of statistical tests on them. He arranged the guesses in order from highest to lowest and graphed them to see if they would form a bell curve. Then, among other things, he added all the contestants' estimates, and calculated the mean of the group's guesses. The number represented, you could say, the collective wisdom of the contestants. "If the crowd were a single person," Surowiecki explains, "that was how much it would have guessed the ox weighed."

Galton thought the average guess of the group would be way off the mark. "After all, according to his belief, mix a few very smart people with some mediocre people and a lot of dumb people, and it seems likely you'd end up with a dumb answer," Surowiecki writes. But Galton was way off the mark. The collective "dumb" crowd had guessed that the ox, after it had been slaughtered and dressed, would weight 1,197 pounds. It ended up weighing 1,198 pounds. In other words, the crowd's judgment was essentially perfect.

Build Measurement Ownership Right at the Start

Those who will be measured must create the measurements. Make sure your entire team buys into your measurement system and goals before you begin using them. The first purpose of the measurement creation process is to help your team understand and agree on what problems must be solved. Make sure your entire team agrees on the problems before you ask them to begin solving them.

When product delivery teams create their own measurement systems, they will "own" and use the measurements. Management's role is to set key product goals. Teams then use these goals as a framework for building their own measurement system. The team then takes its goals and measurement system back to management for their understanding and approval.

Try This as an Experiment

The next time you are in a new restaurant with your friends, build a measurement system for rating the value you receive. First ask what your companions consider the attributes of an excellent restaurant. Almost immediately you will hear categories like affordability, friendliness of service, speed of delivery, cleanliness, atmosphere, quality of food, selection of food, and more. Then ask your friends to individually rate your new restaurant on each of these attributes using a scale of 1 through 10, with 10 being perfection and 1 a non-starter. Ask them also to explain the "why" behind each of their ratings. Make sure they do this without revealing their rating and reasons to each other. Then have each person reveal his or her rating and individually give you their reasons for it. You will be surprised at how consistent their findings are. While we all rate things differently, we generally have the same reasons for doing so.

Countering Measurement Excuses

Excuse #1 – "We are already doing it."

There are three real time metrics that are always used in any project. I always ask folks at my workshops what these "common three" are. In a snap the answer comes back: "schedule, budget and requirements." These three are well known, precisely measurable and easily tracked. We understand them. We are comfortable with them. It always seems we end up tracking what we understand well and can precisely measure, even though these are not the factors that can kill our project.

InnovationCUBE measurement goes well beyond these to answer the questions, "Are we delivering the right value, minimizing risk, and preventing the creation of poor quality and high cost?" CUBE metrics tell you if your business strategy is working while you still have time to make course corrections. What you need, if you are truly going to reach your innovation destination, is a way to leverage measurement for rapid learning. The real power of measurement is not to "prove" but to "improve."

What you must measure, in addition to the "common three," is whether you are on track in *optimizing total value* and *minimizing waste*. Both value and waste are measurable and predictable starting at your early concept stage. You don't have to wait until your idea has hardened into a prototype or detailed drawing. In fact, by then the human mind, battered by a shrinking schedule and evaporating dollars, has little tolerance for change.

Excuse #2 – "Some things can't be measured."

It was Albert Einstein who once remarked, "Not everything that can be measured is worth measuring, and not everything worth measuring can be measured." He was dead right on the first part, but way off the mark on the second. *Everything can be measured.* It really depends on what you mean by "measurement." Einstein's measurement paradigm was constrained by his precise world of physics and mathematics. However, ask an automobile designer, fashion stylist, Olympic judge, art collector or any number of other experts about how they measure success. Be prepared for a deep lesson in the reality of measurement. You may not find a micrometer being used, but you will always find a metric.

Excuse #3 – "Measurement requires data. We don't have data."

There are two worlds of measurement. One is the world we can measure with precision. The other is subjective, the world we know in our minds and experience to be true. Both are important in the world of corporate innovation. And if I had to place my bet on either of them, I would choose the second.

When it comes to front end innovation, relevancy is far more important than accuracy. Precise data is usually not available at the "fuzzy front end" of an innovation effort. However, it is far better to confirm that you are directionally correct, rather than waiting until you have all the precise data you desire. Measure for direction first.

I sometimes encounter managers who believe that a metric is invalid if it cannot be quantified with scientific precision. We must constantly remember what Peter Drucker said so long ago, "Innovation is not a science. Nor is it an art. It is a practice."

In the field of "hard" science we are accustomed to using precise instruments of measurement. We think in two decimal places. We discount anything that is not based on "hard data." Yet, when you really think about innovation, there is seldom any real "hard data" available at the front end of your quest. Innovation begins with theory. We suppose something might be possible and then we pursue it to see if it is so. We move ahead based on very little data, hoping to find validation along the way. We use what is called "soft data." These are suppositions, assumptions and ideas based on what we *feel* might be true. If we have no hard data, we plow ahead anyway, knowing that all the "hard data" of science will eventually be overcome in the future with new discovery.

Excuse #4 – "No group of people can ever agree on the same measurement, so why measure?"

The measurement we are talking about here is a communication tool. It helps share data, ask the unasked questions, and get people thinking from different directions.

Norval Hawkins used "spider" charting, or what we call today "radar" charts, to rate all of Ford's lean product attributes concurrently. Such a chart visually shows in one graphic the size of gaps among a number of both current and future product performance standards. All success factors are measured on the same chart at the same time and the performance of all success factors is shown on one single page. This enables the team to see where the product is "out of balance" and take corrective action. Team members can also simulate "what if" scenarios to gauge the impact of a design change on the product's total success. Results are highly visible, with the "plus" and "minus" factors clearly shown.

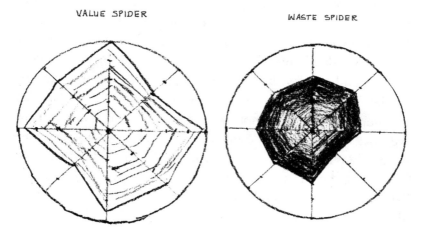

VALUE SPIDER   WASTE SPIDER

From Norval Hawkins' "Rules of Innovation" notebook, 1907.

Innovation is closing the gap.

Norval Hawkins' Rules of Measurement

Hawkins was not your normal accountant. He had a broader perspective when it came to measurement. Jotted in his notebook we find his rules for innovation measurement.

1. First Rule of Measurement. That which gets measured, gets done. We behave as we are measured. Make sure you are measuring what is important, not just what is easy to measure.

2. Rule of Real Time. Measure in "real time," so that you still have time for corrective actions.

3. Rule of the Easy Three. Go well beyond "schedule, budget and technical specs" to measure strategic values, key success factors, and other "drivers" of success.

4. Rule of Intent. Measure primarily to improve, not to "prove." First use measurement to understand.

5. Rule of Simplicity. Fewer is better. Measure the important few, not the trivial many.

6. Rule of Ownership. Never adopt a measurement system, metric or goal without first attempting to make sure all the stakeholders agree it is the correct one.

7. Rule of Relevancy. Relevancy is more important than accuracy. Rough indicators are OK if absolute data is not yet available.

8. Rule of Direction. Measure only to the precision required. Make sure you are headed in the right direction before you begin tracking precise degrees. Rough indicators are OK, especially at the beginning of a project.

9. Rule of Stakeholder Expectations. Surface all measurement expectations, or "pet metrics," as early as possible. Try to make sure everyone is using the same "yardstick."

10. Rule of Uncommon Sense. We often spend most of our time measuring those things we understand well and can measure precisely, even though they are not the elements that are most important for success.

11. Rule of Concurrent Measurement. Measure all key metrics concurrently.Spider charting is a way to do this. Take a "systems view." One measurement goal typically impacts others.

12. Rule of Measurement Alignment. Make sure your metrics are aligned with your customers' goals and your company's goals.

13. Rule of Measurement Iteration. You will never get the right metrics the first time. Be willing to change. Iterate and fine tune.

14. Rule of Knowing Where You Are Now. First baseline where you are now.Understand what you are about to change before you change it. All measurement requires three points: a starting point, a goal, and a progress point.

15. Rule of Totality. Everything can be measured.

16. Rule of Knowledge. Measurement brings knowledge, no matter how coarse the data.

17. Rule of Framework First. Build your measurement system in an orderly way: Measurement system framework, metrics, goals and baselines. Do these steps in that order.

18. Rule of Visual Metrics. Use highly visual metrics which can be grasped quickly; they will have high impact.

19. Rule of the Clock. Set time limits, no matter how small the measurement task. Time measurement reduces wasteful meetings.

20. Rule of Value Measurement. Don't ever measure anything where the benefit is less than the cost of capturing the data.

21. Rule of Guidance. Create a measurement system that guides you, not one that rates you!

PART III

The Rest of the Story

CHAPTER EIGHT

Henry Ford Stumbles

Fatal Flaw: You fail to convert the Rules into a sustainable process.

U p to 1915, the story of Henry Ford and the Model T was a happy and extraordinary one. The Ford Motor Company had written one of the most brilliant chapters of industrial history. A few men, starting with a pittance, had in those years built an industrial empire on the foundation of the Model T. They had done more than all other manufacturers combined to make the automobile a "democratic possession." On May 31, 1921, the Ford Motor Company turned out car No. 5,000,000. The Model T had changed, at least in the mind of the marketplace, very little. Ford begrudgingly added a starter — and a battery to start it. But he was proud to say that the changes brought to the Model T were in the "making and not through any change in the basic design, which I take to be an important fact that, given a good idea to start with, it is better to concentrate on perfecting it than to hunt around for a new idea. One idea at a time is about as much as anyone can handle."

Two-thirds of all the automobiles registered in the United States in 1923-1924 were Ford Model Ts, fondly called *Tin Lizzies*. By then the world did not *Want* cars — it had already come to *need* them. The Model T held a virtual monopoly on America's new household necessity.

No Sustainable Innovation Process

Yet Ford had no sustainable, systematic innovation process in place to replicate his early success. There was no corporate architecture to perpetuate his early genius. Ford Motor began to slide into an innovation rut, lost market share and almost faced oblivion. Ironically, the leaders who had worked with him on the design of the Model T left Ford and became Henry Ford's toughest competitors at General Motors.

Over its nineteen-year life span, changes in the Model T design were barely perceptible. Henry Ford had mandated it so. To Ford, the Tin Lizzie was perfection. The company's efforts went into producing more cars at lower cost. A humorist of the time noted, "Two flies can manufacture 48,876,532,154 new flies in six months. But they have nothing on those two Ford factories."

But Ford's green banana of 1908, when the first of the Model T autos were shipped, was definitely turning into one with growing flecks of brown. The attraction of the Model T was beginning to dim. In its place was emerging a new green banana.

How Ford Stumbled

It was Henry Ford himself who said, "Businessmen go down with their businesses because they like the old way so well they cannot bring themselves to change. One sees them all about — men who do not know that yesterday is past, and who woke up this morning with their last year's ideas."

In his early automotive days, Henry Ford was superb leader. The design of the Model T, the development of the moving assembly line, and the genius of doubling the wage of his employees to five dollars a day, all had the imprint of his personal leadership. But then things began to slip.His personal leadership shadow was not long enough to reach across his widening enterprise.

Henry Ford had applied Norval Hawkins' *Rules of Innovation* to build his automobile empire. He then broke Hawkins' rules one after another. The story of Ford is a sobering account of how success can carry within it the seeds of its own destruction. Every innovation begins to degrade the day it is launched... many times even before it is introduced to the marketplace.

How Ford Failed Step-by-Step

Ford failed to recognize that the public was beginning to want more than his strictly utilitarian Model T. He had mandated all Model Ts would be painted black to increase production schedules. Chevrolet moved ahead by offering multiple colors and features not offered in the Model T.

Ford became stuck in the past. While the Ford Motor Company consistently improved its production techniques, the car itself was unchanged. Henry Ford set no new goals for it nor even entertained the idea of a multiple car line. And he forgot his own rules.

Ford even violated his first rule about building a car under one roof. He had once said, "The most economical manufacturing of the future will be that in which the whole of an article is not made under one roof—unless, of course, it be a very simple article. The modern — or better, the future — method is to have each part made where it may best be made and then assemble the parts into a complete unit at the points of consumption. That is the method we are now following and expect to extend." He built the River Rouge, Michigan production complex and integrated all manufacturing from the growing of his own trees to the disposal of his own waste. Ford Motor fell into a "competency trap."

Remember the Three Sharks of the changing Marketplace, Technology and Competition. The marketplace is always changing, wanting yet something else that is new, while not adapting to changing technology can be suicide. A prime example is Henry Ford's refusal to include an electric starter and battery. Consumers wanted the convenience, and the technology was available, but Ford actually *created* competitors by encouraging his top managers to leave Ford Motor. They took their Ford experience and quickly applied it at General Motors and Chrysler Corporation.Norval Hawkins was one of them. In 1919, Hawkins left Ford Motor for General Motors.

CHAPTER NINE

The InnovationCUBE©

T he InnovationCUBE© integrates all the Rules in this book into a systematic process you can use to lead any project team to success. It is contained on a USB "stick." You can use the CUBE as an "innovation roadmap" for any kind of project.

The CUBE is not a box of old tools packaged in a new way. It is a *process* you can apply to optimize your chances for success. Its principles, rules and steps have been used intuitively for centuries. All the CUBE does is to bring these together in a seamless, step-by-step manner.

Reduce Cost by Thirty-five Percent or More

A leading Navy shipbuilder is now using the CUBE to reduce costs by thirty-five to seventy-five percent. Project teams use the Rule "Stop Waste Before It Starts," with its description of the

"*Evil Ings*" (Chapter 4) as a lens to identify the primary causes of poor quality and high cost in the present ship design. They use the rule "Measure to Learn," (Chapter 7) to quantitatively baseline the existing design. Their objective is "fire prevention, not fire fighting." New designs are evaluated against this baseline. This same technique can be used with transactional processes.

Validate Product Strategy

A leading software company wanted to validate its strategy for reaching emerging markets. Were they on the right track? The project team elevated their thinking from product requirements to the level of the eight *Values* they considered the criteria for success. Using the rule "Discover Hidden Wants" (Chapter 3), they found that the *Values* they thought were correct were way off base.

Look for New Market Opportunities

A banking company wanted to seek new market opportunities. They used the InnovationCUBE to benchmark their competitors and looked for "gaps" that they could fill with new services. The result: a forty percent increase in business in one service line.

Use Measurement to Compare Alternatives

A senior management team was having difficulty making decisions between alternate portfolio strategies. They had no way to compare several approaches. They used the measurement features of the InnovationCUBE (see Chapter 7) to create a system for quantitatively arriving at the best business decision.

For more information on the InnovationCUBE, contact the Institute for Lean Innovation, 9 French Outpost, Mackinac Island, Michigan 49757, or visit www.InnovationCube.com.

A Guide to Current
Innovation Terminology

Blue Ocean Strategy. "Blue oceans" are uncontested market space, far removed from the "red oceans" of hotly contested market space. Term comes from the book *Blue Ocean Strategy* by W. Chan Kim and Renee Mauborgne, who maintain that market leaders succeed *not* by battling competitors, but by finding uncontested markets ripe for growth.

Collaborative Innovation. Term used to describe involvement of all stakeholders beginning with a project's concept phase. This term is also used to describe collaboration with an outside supplier or another company, either on a formal or informal basis.

Competency Trap. Potential for an organization to become so competent in its product and sphere of business that it falls prey to thinking its skills and expertise will last forever. The company loses the desire and flexibility to seek new markets and skills. Ford became so invested, both from the standpoint of culture and capital, that it was unable to adjust quickly to the changing automobile market.

Critical to Quality (CTQ). Term typically used to describe technical characteristics of a product or service offering.

Design for Manufacture & Assembly (DFMA). Technique of focusing on the manufacturing and assembly processes to reduce cost and improve quality. Term reportedly first used during the design of the Ford Taurus.

Design for Six Sigma (DFSS). Well known process for using Six Sigma techniques to improve a product in the early development stage.

Design for X (DFX). Term used to describe the practice of designing for all-important attributes beyond manufacturability and assemble-ability to include "*Ilities*" such as post process maintainability, serviceability, install-ability and others.

Disruptive Innovation. Concept that new organizations can use relatively simple, convenient, and low-cost innovations to create growth and triumph over established companies. Author Clayton M. Christensen put forth this theory in his book *The Innovator's Dilemma.* Existing companies have a high probability of beating back new entrants when the contest is about *sustaining innovations*, but are relatively weak in reorganizing themselves to meet radically new threats.

Divergence Strategy. Concept that says if you want to create a powerful new brand, you should look for ways that your product or service can *diverge* from an existing category. This strategy is expounded by authors Al and Laura Ries, in their book *The Origin of Brands,* who argue that the best way to build a brand is not by going after an existing category, but by creating a new category you can be first in.

Early Adopters. Term used to describe consumers, including corporations, that lead the marketplace in the adoption of new products or processes. Six Sigma early adopters were Motorola, ABB and

General Electric. Lean manufacturing, or Toyota Production System, early adopters included automotive companies such as Ford Motor. However, to call Ford Motor an "early adopter" of lean manufacturing would be erroneous, since lean manufacturing in the true sense of the word began in the Ford Piquette manufacturing plant and was fine-tuned in the Ford Highland Park plants.

Empathetic Design. An idea-generating technique where innovators systematically observe how people use existing products and services in their own environments.

Eureka Moment. Concept that a "Eureka Moment" or *Insight* is the combination of three "sights" or skills: *Foresight, Hindsight* and *Outsight.*

Evil Ings. Processes that create potential for quality loss, high cost, and slow time to market. The term is used to describe no-value processes such as inspect-ing, fix-ing, repair-ing, and other non-value-added tasks. Concept of *ing* elimination based on fact that all products, services and processes are combinations of tasks, or *ings.*

Fuzzy Front End. Term used to describe the conceptual phase of a project when goals, data and problems are all unclear. First used more than twenty years ago to emphasize the importance of conceptual thinking.

Idea Funnel. A metaphor for the practice of "pouring in" a large volume of ideas and then using a screening process to sort the good from the bad.

Ideation. Most times used to describe the "front end" of a project when ideas are used to create solutions.

Ilities. Attributes that define positive characteristics of a product, service, transactional process, or strategy. Sometimes referred to as Key Success Factors (KSFs) or Critical To Quality (CTQs).

Ings. Tasks used to create value, products or services from all Four Domains, such as designing, manufacturing, supplying, marketing, and all the sub-tasks of these.

Innovation Process Management. Organizations often try to start new-growth businesses using processes that were designed to make their mainstream business run effectively. *Innovation Process Management* is a growing body of knowledge and methods focused on making innovation understandable, repeatable, scaleable and measurable.

Insights. Insights are created at the intersection of what this book calls the "Three Sights." *Foresight* is the ability to "see" unmet *Wants* of the present or future. *Hindsight* is the ability to draw upon the knowledge and experience of the past. And *Outsight* is the ability to tap into different disciplines outside the scope of one's normal experience. These three skills are learnable, repeatable and manageable within the modern corporate environment. See also *"Eureka Moment."*

Lead Users. Customers who are the first to adopt a new product, service or system. Lead users typically modify offerings to customize to their wants. Observing these lead users and learning why and how they adapt a new offering give clues as to how to improve an innovation.

Lean Innovation. Term given to a new product, process or business system that focuses on delivering new value with far less cost, better quality and faster time to market.

Lean Sigma. Term used by companies in the process of merging the discipline of lean thinking with the quality mindset of Six Sigma.

Medici Effect. Describes the concept that world-changing discoveries will come from the *intersections* of disciplines, not from within them. Term was popularized in the book by Frans Johansson, *The Medici Effect.* It refers to a remarkable burst of creativity in fifteenth-century Renaissance Italy, funded by the wealthy Medici family in Florence, who brought together a diverse combination of cultures, ideas and concepts.

Open Innovation. The practice of gaining competitive advantage from leveraging the discoveries of others. Term used by Henry Chesbrough in his book *Open Innovation*. He convincingly argues that successful businesses today must organize themselves to use both internal and external sources of ideas in order to stay competitive. See Chapter 5 and the rule "Search Outside Your Shell" for how to systematically do this.

Organic Growth. Term used to describe top line growth from internal resources, primarily from new, innovative products and services. The opposite of *Organic Growth* is growth through acquisitions. Wall Street is increasingly using *Organic Growth* as a measure of a company's future potential.

Overshot. Word used to describe products that deliver more than the marketplace wants or is willing to pay for. The opposite is "Undershot," not providing the marketplace enough value.

Phase Gate. A linear process for assuring certain criteria are met at specific times during a project development effort. The intent is to surface problems or uncertainties before moving to the next stage of development.

Process Excellence. The practice of seeing business as a series of processes, each of which can be constantly improved. For example, Henry Ford considered his company to be the master of manufacturing process excellence. He saw Model T production as tapping into the "natural code" of process excellence.

Real Time Metrics. Sometimes referred to as "Predictive Measurements." This term refers to metrics that are used to track a project's potential for success from the early concept phase. These typically go beyond the "Common Three Metrics" of schedule, performance and budget.

S-Curve. Plot of typical product showing acceptance on the Y axis and time on the X axis with acceptance flattening and then dropping as new technology emerges.

Set Based Design. The practice of developing several solutions, or "sets," at the same time. The objective is to look at many approaches before making the decision to arrive at a final approach. The opposite of this is "point based design," where a decision is made early to focus on one approach. The term "set based design" is attributed to Toyota's design approach in the automotive industry.

Spider Charting. Use of "radar" or "spider" charts to measure project progress against baseline metrics.

Sweet Spot. Analogy taken from golf, where the club head strikes the ball with the correct timing, speed and angle. You have struck the innovation "sweet spot" when you find and match a marketplace *Want* with a timely solution (*Thing*) through the use of your *Insight.*

Systematic Corporate Innovation. The process of systematically replicating how innovations can occur within, and outside, an organization structure. The term "systematic innovation" was first used by Peter F. Drucker in his book, *Innovation and Entrepreneurship,* published in 1985. Drucker contended that innovation is neither a science nor an art, but a practice that can be organized as day-to-day work.

Technology Brokering. The process of creating networks of ideas between separate industries in order to create breakthrough innovations. The term was coined by Andrew Hargadon in his book *How Breakthroughs Happen.*

Technology Transfer. Term used to describe movement of one technology to another, sometimes completely new field. An example is the technology transfer of email order patterns from the

Amazon.com commercial goods domain to the technology of automotive model ordering prediction.

Things. Term used in this book to encourage the reader to see the total universe of innovation solutions, including but not restricted to technologies, processes, and policies. All *Things* are the integration of the "tasks to be done" to deliver a value. "Tasks to be done" are all the processes required to create "any-thing." Thus, *Things* are the sum of their "*-ings,*" or total tasks.

Three Sharks. Idea that three forces of constant change begin to obsolete any innovation the day it is introduced. These Three Sharks, as described in Bart Huthwaite's book *The Lean Design Solution,* are (1) The marketplace's constant appetite for something new and better; (2) New technology; and (3) New competitive threats. Knowing the way these Three Sharks are headed can help you develop both a defense and an offensive strategy for dealing with them.

Tipping Point. Term popularized and applied to daily life by Malcolm Gladwell's book, *The Tipping Point: How Little Things Can Make a Big Difference.* The concept has been applied to any process in which, beyond a certain point, the rate at which the process moves forward increases dramatically. The explosive growth of a new style is used as a common example.

TRIZ. A Russian acronym for "*Teoriya Resheniya Izobretatelskikh Zadatch*" a theory of solving inventive problems developed by Genrich Altshuller, a Russian, and his colleagues since 1946. TRIZ is an engineering tool that uses an algorithmic approach to the invention of new systems and the refinement of old systems.

Undershot. Word coined by author Clayton Christensen to describe a product or service offering that is below a customer's expectations. See "Sweet Spot" for a definition of the ideal value customers seek.

Wants. Term used in this book to prompt the reader to think beyond marketplace "needs" to discover untapped or unmet desires. The most profitable innovation successes come from discovering, meeting, and exciting new *Wants* with the hope of converting them to "needs." At first we don't "need" Starbucks Coffee, laptop computers, flat screen television, and cell phones.

Further Reading for Innovation Leaders

Innovation Management

Anderson, Chris. *The Long Tail: Why the Future of Business is Selling Less of More.* New York: Hyperon Press, 2006.

The theory of Anderson's Long Tail is that our economy is increasingly shifting away from a focus on a relatively small number of mainstream hits and markets at the head of the demand curve and moving towards a large number of niches at the tail. The long tail of small niches, Anderson maintains, is equal to the large head of the demand curve.

Christensen, Clayton M. *The Innovator's Dilemma.* Boston: Harvard Business School Press, 1997.

Christensen, Clayton M. and Michael E. Raynor. *The Innovator's Solution: Creating and Sustaining Successful Growth.* Boston: Harvard Business School Press, 2003.

Conner, Clifford D. *A People's History of Science: Miners, Midwives, and "Low Mechanicks."* New York: Nation Books, 2005.

Davila, Tony, Marc J. Epstein, and Robert Shelton. *Making Innovation Work: How to Manage It, Measure It, and Profit from It.* Upper Saddle River, NJ: Wharton School Publishing, 2006.

Drucker, Peter F. *Innovation and Entrepreneurship: Practice and Principles.* New York: HarperBusiness, 1986.

Edersheim, Elizabeth Haas *The Definitive Drucker.* New York: McGraw-Hill, 2007.

Gourville, John T. "Eager Sellers and Stony Buyers: Understanding the Psychology of New-Product Adoption." *Harvard Business Review,* June 2006, pp. 98-106.

Hargadon, Andrew. *How Breakthroughs Happen: The Surprising Truth About How Companies Innovate.* Boston: Harvard Business School Press, 2003.

Andrew Hargadon reaches back into history to argue that revolutionary innovations do not result from flashes of brilliance by one inventor or organization. He says innovation is really about creatively recombining ideas, people, and objects from past technologies in ways that spare new technological revolutions. He coins the term "technology brokering" to explain the networked, social nature of the innovation process.

Henderson, Bruce A. and Jorge L. Larco. *Lean Transformation: How to Change Your Business into a Lean Enterprise.* Richmond, VA: The Oaklea Press, 2000.

This is an easy-to-read, practical guide for applying the Toyota lean principles. The authors go beyond the manufacturing floor to show how lean thinking can be applied to transactional processes.

Huthwaite, Bart. *The Lean Design Solution: A Practical Guide to Streamlining Product Design and Development.* Mackinac Island, MI: Institute for Lean Innovation Press, 2004.

Immelt, Jeff. "Growth as a Process," Interview of Jeff Immelt, CEO of General Electric.*Harvard Business Review,* June 2006, pp. 60-70.

Johansson, Frans. *The Medici Effect: Breakthrough Insights at the Intersection of Ideas, Concepts, and Cultures.* Boston: Harvard Business School Press, 2004.

Kim, W. Chan and Renee Mauborgne. *Blue Ocean Strategy: How to Create Uncontested Market Space and Make the Competition Irrelevant.* Boston: Harvard Business School Press, 2005.

The authors challenge companies to break out of the "red ocean" of bloody competition by creating uncontested market space that makes competition irrelevant. Instead of dividing up existing demand and benchmarking competitors, Kim and Mauborgne argue for developing a "blue ocean" strategy for growing demand and breaking away from competitors.

Koestler, Arthur. *The Act of Creation.* London: Penguin Books, 1989.

Koestler affirms that all humans have the capacity for creative activity, frequently suppressed by the automatic routines of thought and behavior of daily life. He examines what he terms "bisociative" thinking — the linking of different planes of reference by the mind to make the creative leap that gives rise to new and startling perceptions and glimpses of reality.

Minsky, Marvin. *Thee Society of the Mind.* New York: Simon & Schuster, 1988.

A brilliant series of essays on how the mind works. Minsky is considered one of the fathers of computer science and is a co-founder of the Artificial Intelligence Laboratory at MIT. The book was named a "Notable Book of the Year" by the *New York Times Book Review.*

Moore, Geoffrey A. *Dealing with Darwin: How Great Companies Evolution.* New York: Penguin, 2005.

Nadler Gerald and Shozo Hibino. *Breakthrough Thinking: Why We Must Change the Way We Solve Problems, and the Seven Principles to Achieve This.* Rockland, CA: Prima Publishing and Communications, 1990.

This book describes how to innovate using what the authors call the solution-after-next principle. They say that an innovative solution cannot be bolted in place like a piece of equipment but should be introduced as part of an ongoing process. Due thought must be given to transitional phases. They point out that before we do something new, we must undo something old. The authors' solution-after-next approach says that the solution-after-next is simply what your competitor comes up with after you think you've "finished" a project. This problem arises due to neglecting to look beyond the immediate problem and the solution-after-next.

Ries, Al and Laura. *The Origin of Brands: How Product Evolution Creates Endless Possibilities for New Brands.* New York: Harper Collins Publishers, 2004.

Rubenstein, Moshe F. and Iris R. Firstenberg. *The Minding Organization: Bring the Future to the Present and Turn Creative Ideas into Business Solutions.* New York: John Wiley & Sons, 1999.

The authors use the metaphor of the human body to describe how organizations must act in an increasingly chaotic world. Every individual has a mind, used to perceive, judge, make decisions and adapt. The analogy between the human and the "minding organization" begins with the mind. A minding organization must have a sense of purpose and articulate the sense of purpose throughout. The minding organization must focus on the future it wants to create. The book is mainly directed toward senior leadership.

Schwartz, Evan I. *Juice: The Creative Fuel that Drives World-Class Inventors.* Boston: Harvard Business School Press, 2004.

Seidensticker, Bob. *Future Hype: The Myths of Technology Change*. San Francisco: Barrett-Keohler Publishers, Inc., 2006.

One of my favorite books. I call the author a "technology bubble buster." Seidensticker will recalibrate your thinking by looking at how technology change really happens. The popular perception of modern technology is inflated and out of step with reality. We overestimate the importance of new and exciting inventions and underestimate those we've grown up with. He gives valuable lessons in how technology emerges in fits and starts and how it can "bite back."

Surowiecki, James. *The Wisdom of Crowds*. New York: Anchor Books, 2004.

Surowiecki explores a deceptively simple idea: Large groups of people are *smarter* than an elite few, no matter how brilliant. They are better at solving problems, fostering innovation, coming to wise decisions and even predicting the future. The author is a staff writer at *The New Yorker* where writes the business column, "The Financial Page."

Trout, Jack with Steve Rivkin. *Differentiate or Die: Survival in Our Era of Killer Competition*. New York: John Wiley & Sons, 2000.

The average supermarket has 40,000 brands on its shelves, and car shoppers can wander through the showrooms of more than 20,000 automobile makers; in today's ultra-competitive world the amount of opportunities is dazzling. Trout points out that any product or service has to clearly differentiate itself if it wants a chance to survive. Just being creative is not enough, and just being innovative is not enough. What a product or service has to do is to differentiate itself.

Utterback, James M. *Mastering the Dynamics of Innovation: How Companies Can Seize Opportunities in the Face of Technological Change*. Boston: Harvard Business School Press, 1994.

Vogel, Craig M., Jonathan Cagan, and Peter Boatwright. *The Design of Things to Come: How Ordinary People Create Extraordinary Products.*Upper Saddle River, NJ: Wharton School Publishing, 2005.

Williams, Roy H. *The Wizard of Ads.* Austin, TX: Bard Press, 1998.

Womack, James P., Daniel T. Jones, and Daniel Roos,. *The Machine That Changed the World.* New York: Macmillan Publishing Company, 1990.

History of Innovation

Asimov, Isaac. *Asimov's Biographical Encyclopedia of Science and Technology: The Lives and Achievements of 1510 Great Scientists from Ancient Times to the Present Chronologically Arranged.* New York: Doubleday & Company, 1982.

Burke, James, *Connections.* Boston: Little, Brown and Company, 1978.

This is a brilliant recounting of the ideas, inventions, and coincidences that have led to the great technological achievements of today. This book was the basis for a series on PBS in 1979. Burke untangles the pattern of interconnecting events that gave rise to the development of science.

Conner, Clifford D. *A People's History of Science: Miners, Midwives, and "Low Mechaniks."* New York: Nation Books, 2005.

Conner presents startling new historical data that shows that science has always been a collective endeavor with many people playing a role.

Heilbron, J.L. *Elements of Early Modern Physics.* Berkeley: University of California Press, 1982.

Levy, Joel. *Really Useful: The Origins of Everyday Things.* Willowdale, Ontario: Firefly Books Ltd., 2002.

Having explored centuries of innovation, Levy points out that the real driving forces behind invention and innovation are social and cultural ones. He emphasizes that nothing is really new; it's all simply re-combinations and modifications of things from the past.

Pinati, Charles. *Extraordinary Origins of Everyday Things.* New York: Harper & Row, 1987.

This book is for those who want examples of how innovation *really happened.* In its 463 pages, Panati reveals innovation through examplesthat include the invention of the pretzel, the hoola hoop, the magazine, the band aid, buttons, zippers, and Velcro. All of these stories have a common theme that resonates directly with the fundamental dynamics of innovation.

Creativity Tools

Ritter, Diane and Michael Brassard. *The Creativity Tools Memory Jogger.* Salem, NH: Goal/QPC, 2006.

Maxwell, John C. *Thinking for a Change: 11 Ways Highly Successful People Approach Life and Work.* New York: Warner Business Books, 2003.

Michalko, Michael. *Cracking Creativity: The Secrets of Creative Genius.* Berkeley, CA: Ten Speed Press, 2001.

Each chapter gives a different creative thinking strategy to help you reorganize the way you think. A well-written book for building your idea generating horsepower.

Papageorge, Andrew. *GoInnovate! A Practical Guide to Swift, Continual and Effective Innovation.* San Diego, CA: GoInnovate! Publishing, 2004.

Papageorge tells us that everyone in an organization has two jobs: one is working on the business of today while the other is innovating, or creating, the business of tomorrow.

Root-Bernstein, Robert and Michele. *Sparks of Genius: The 13 Thinking Tools of the World's Most Creative People.* Boston: Houghton Mifflin Company, 2001.

Stauffer, Dennis *Thinking Clockwise: A Field Guide for the Innovative Leader.* Minneapolis, MN: MinneApplePress, 2005

Dennis Stauffer designed this small book (104 pages) to be a simple, user-friendly guide to both individual innovative thinking as well as leadership. He uses the clock as his metaphor. Clockwise thinking is positive, self-reinforcing and creative.

Usher, Abbot Payson. *A History of Mechanical Inventions,* revised ed. Dover Publications, 1954.

Von Oech, Roger. *A Kick in the Seat of the Pants.* New York: Harper & Row, 1986.

A classic that is a "must" for your bookshelf.

Von Oech, Roger. *A Whack on the Side of the Head,* revised ed. New York: Warner Books, 1990

Another of von Oech's classics. A deep mine of ways to challenge "stuck in a rut" thinking.

Von Oech, Roger. *Expect the Unexpected or You Won't Find It.* San Francisco, CA: Berrett-Koehler Publishers, 2002.

A creativity tool based on the thinking of Heraclitus, a Greek philosopher who lived 2,500 years ago.

Watson, Peter. *Ideas: A History of Thought and Invention, from Fire to Freud.* New York: HarperCollins Publishers, 2005.

This book is a grand sweep of the ideas that have shaped our civilization, from deep antiquity to the present day. Watson's monumental work of 822 pages clearly reveals how all ideas are in some way linked and why "new ideas" may not be so new. This book is excellent for your reference shelf but would be a tough read on your next flight.

Henry Ford

Bryan, Ford R. *Beyond the Model T: The Other Ventures of Henry Ford,*revised ed. Detroit, MI: Wayne State University Press, 1997.

Bryan, Ford R. *Henry's Attic: Some Fascinating Gifts to Henry Ford and His Museum.* Detroit, MI: Wayne State University Press, 2006.

Bryan, Ford R. *Henry's Lieutenants.* Detroit, MI: Wayne State University Press, 1993.

This book is a treasure trove of biographical sketches of the people who contributed to Henry Ford's fame. The author obtained much of the material from the oral reminiscences of the subjects themselves.

Brinkley, Douglas. *Wheels for the World: Henry Ford, His Company, and a Century of Progress, 1903-2003.* New York: Penguin Books, 2003.

One of the best sources for anyone wanting to grasp the historical scope of the Ford Motor Company. This 858-page history was published in 2003, the year the Ford Motor Company celebrated its one hundredth anniversary. I agree with the review by *Business Week*, which called it "A meticulously annotated, highly readable, and engrossing book... Brinkley's warts-and-all account of Henry Ford and his company is a winner...T o his credit, Brinkley deftly and even-handedly captures many of the contradictions of the brilliant inventor, manufacturing genius, abysmal businessman, and stubborn iconoclast who nearly ran his company into the ground several times."

Brown, Mark Graham. *Keeping Score: Using the Right Metrics to Drive World-Class Performance.* New York: Quality Resources, 1996.

Crabb, Richard. *Birth of a Giant: The Men and Incidents That Gave America the Motorcar.* New York: Chilton Book Company, 1969.

A well written narrative on the men who shaped the early auto industry in America. Crabb cracks open the minds of such auto greats as Duryea, Olds, Chrevrolet, Durant, Leland, the Dodge Brothers, Chrysler and Henry Ford.

Edison & Ford Quote Book, The. Edison and Ford Winter Estates, Ft. Myers, Florida, 2004.

Ford, Henry and Samuel Crowther. *My Life and Work.* Kessinger Publishing Rare Reprints.

Head, Jeanine M. and William S. Pretzer. *Henry Ford: A Pictorial Biography.* Dearborn, MI: Henry Ford Museum & Greenfield Village, 1990.

Lacy, Robert. *Ford: The Men and the Machine.* Boston: Little, Brown and Company, 1986."

Covers four generations of Fords with all the characters that surrounded them.

*Managing Creativity and Innovation.*Harvard Business Essentials. Boston: Harvard Business School Publishing, 2003.

Nevins, Allan. *Ford: The Times, The Man, The Company.* New York: Charles Scribner's Sons, 1954.

A "must read" classic that sets the tale of Henry Ford within the context of his times. Nevins seamlessly stitches the man into his surrounding environment to help the reader better understand what drove his leadership. This 668-page well-written history covers the formative Ford Motor Car period from its founding in 1903 to 1915, when Ford, the man, began to shift his leadership style to that of an autocrat.

Olson, Sidney. *Young Henry Ford: A Picture History of the First Forty Years,* revised ed. Detroit, MI: Wayne State University Press, 1997.

A visual look at Henry Ford from his birth up to the formation of the Ford Motor Company in 1903.

About the Author

Bart Huthwaite, Sr. is a world renowned expert in innovation leadership. He is the founder of the Institute for Lean Innovation and the thought leader in the emerging business process known as **"Sustainable Corporate Innovation."** This is a method for giving managers the knowledge to make corporate innovation understandable, repeatable and, very importantly, measurable.

Bart Huthwaite, Sr.

Founder, Institute for Lean Innovation

Lecturer, University of Michigan College of Engineering

The **University of Michigan College of Engineering** has adopted Huthwaite's work and is now offering courses in it at the U-M Center for Professional Development.

Huthwaite has mentored managers and teams in corporate innovation worldwide at more than **1,000 companies** over the past **twenty-five years.**

His last book, ***The Lean Design Solution: A Practical Guide to Streamlining Product Design and Development,*** is available on Amazon.com. Huthwaite is also the author of ***Engineering Change,*** a "how to" guide for re-inventing your product development process, ***Strategic Design,*** a step-by-step road map for product teams, and ***Lean Innovation User's Guide.***

Institute for Lean Innovation Services

On-Site Workshops: One, two, or three-day on-site workshops for 25-30 participants based on the Rules and InnovationCUBE© described in this book. All workshops are "hands on" with participants working on their own products, processes or services. The Institute has trained over 300,000 people at more than 1,000 leading companies. This workshop is excellent for launching project teams. All workshops are customized for your product and industry and are guaranteed to deliver immediate results.

Needs Assessment: In-depth reviews of your organization's innovation capability, followed by a "no holds barred" presentation to your senior executives on how your company compares to others.

Consulting & Coaching: On-site leadership coaching and consulting in how to build a stronger innovation culture.

Institute for Lean Innovation
P.O. Box 1999
9 French Outpost
Mackinac Island, Michigan 49757
www.innovationcube.com
906-847-6094